Forestry Canada was formerly known as the Canadian Forestry Service. This change of name identifies with the new federal Department of Forestry which was created in 1988.

Cover photo: Spruce budworm

Insects of Eastern Spruces, Fir and Hemlock

A.H. Rose and
O.H. Lindquist

Great Lakes Forestry Centre
Sault Ste. Marie, Ontario

Forestry Canada
Forestry Technical Report 23
Ottawa, 1989

Available in Canada through
Authorized Bookstore Agents and
other bookstores or by mail from:

Canadian Government Publishing
Centre, Supply and Services Canada
Ottawa, Canada K1A 0S9

Catalogue No. Fo64-23/1985E
ISBN 0-660-11818-1

First Edition 1977
Revised 1985
Reprinted 1989 with minor corrections

Cette publication est également
disponible en français sous le titre
Insectes des épinettes, du sapin et de la pruche de l'est du Canada.

Printed in Canada

The other Forestry Canada
handbooks in this series include:

Insects of Eastern Pines
Insectes des pins de l'est du Canada

Insects of Eastern Larch, Cedar and Juniper
Insectes du mélèze, du thuya et du genévrier de l'est du Canada

Insects of Eastern Hardwood Trees
Insectes des feuillus de l'est du Canada

Abstract

This handbook deals with insects feeding on spruces, balsam fir, and hemlock in Canada from the Rockies to the Atlantic. Seventy-five species or species groups causing damage or which are commonly found on these trees are covered. Color illustrations aid in the identification of insects and trees. Flow chart keys help identify the insect and the injury found. Included are biological sketches of causal insects, illustrations of life stages, and the injury caused. Control is discussed along with the required type and timing of applications of pesticides. Injury by birds, mammals, mites and other agents is also included.

Résumé

Le présent manuel traite des insectes qui se nourrissent à même l'épinette, le sapin baumier et la pruche au Canada, depuis les Rocheuses jusqu'à l'Atlantique. L'auteur traite des espèces ou groupes d'espèces qui causent des dégâts ou qu'on retrouve communément sur les arbres. Les illustrations en couleurs aident à identifier arbres et insectes. Des diagrammes clés servent à identifier l'insecte et les blessures qu'il cause. Le lecteur y trouvera des croquis biologiques des insectes nuisibles, des illustrations montrant les divers stades de vie des insectes et aussi les blessures causées aux arbres. Il est question de répression et de pesticide à appliquer et quand l'appliquer. Finalement, on retrouve des illustrations de blessures causées par les oiseaux, les mammifères, les acariens et autres agents destructeurs.

Preface

This handbook, the second in a series, was made possible by the far-sighted vision of some outstanding Canadian entomologists, who conceived, nurtured and encouraged the development of a national survey organization that has over the years acquired a large body of knowledge and expertise on forest insects. Valuable contributions were made by other individuals both inside the Forest Insect and Disease Survey and outside. Among the latter, particular mention should be made of specialists at Canadian Forestry Service Centres across the country and national taxonomists at the Biosystematics Research Institute, Agriculture Canada, Ottawa. A dedicated and inquisitive field staff hastened the solution of many identification problems and obtained much of the material for the illustrations. This handbook, then, is the product of the combined efforts of many people. We, the authors, have accepted the challenge of preparing and presenting it to you.

Contents

White spruce

Black spruce

Red spruce

Norway spruce

Colorado or blue spruce

Balsam fir

Eastern hemlock

Ground hemlock (actually, a dwarf yew)

Introduction

This is the second in a series of handbooks dealing with the identification and biology of important insects of conifers east of the Rocky Mountains. The first in the series, *Insects of Eastern Pines*, was published in 1973. This handbook treats the insects feeding on the various spruces, balsam fir and hemlock.

Spruce, fir and hemlock rank first, second and fifth, respectively, in terms of volume in the coniferous forests of eastern Canada. The naturally occurring species of spruce – white, black and red – are very important pulpwood and lumber species. The introduced species – Colorado and Norway – are commonly planted ornamentals. Balsam fir is an important pulpwood species and in the Atlantic Provinces is a major Christmas tree species. The primary use of eastern hemlock is in construction; ground hemlock (Canada yew) is of no commercial importance.

Some of Canada's major insect pests periodically attack the spruce-fir-hemlock group of trees with results that are economically catastrophic. Furthermore, the appearance of trees is often adversely affected by many other insects. For these reasons there is a need for a handbook that will permit ready identification of the pests causing injury and indicate the consequences of their feeding as a basis for responsible control decisions.

Data for this handbook were drawn in large part from 30 years of records kept by the Forest Insect and Disease Survey, Great Lakes Forest Research Centre, Sault Ste. Marie, Ontario. Further material and information were contributed by other forest research centres of the Canadian Forestry Service and additional biological information on many of the insect species was drawn from the entomological literature. About 400 insect species have been found on spruce, fir and hemlock, but the majority do not cause serious injury. About 75 species or species groups are known to have caused serious damage in eastern Canada. These were selected for inclusion in this handbook and have been grouped on the basis of the part of the tree on which the insect or injury occurs as shown in the table of contents. After the damaged part has been determined the reader is led through flow chart keys to an identification box and a page in the text where the identity of the insect can be confirmed by coloured illustrations, indicated by □. Biological information is also provided on the species involved.

If after using this handbook the reader is still doubtful about the identity or importance of any insects occurring in large numbers on trees or shrubs, he should send a representative sample of living specimens and the damage caused by them to the appropriate Forest Insect and Disease Survey Unit at the Canadian Forestry Service research centre serving each of the areas indicated on the map, page 12.

Injury

Injury to trees can be caused by such varied factors as climate, microscopic pathological organisms, tiny mites, insects of all sizes, birds and mammals. Man himself often causes injury by mechanical means or by interfering

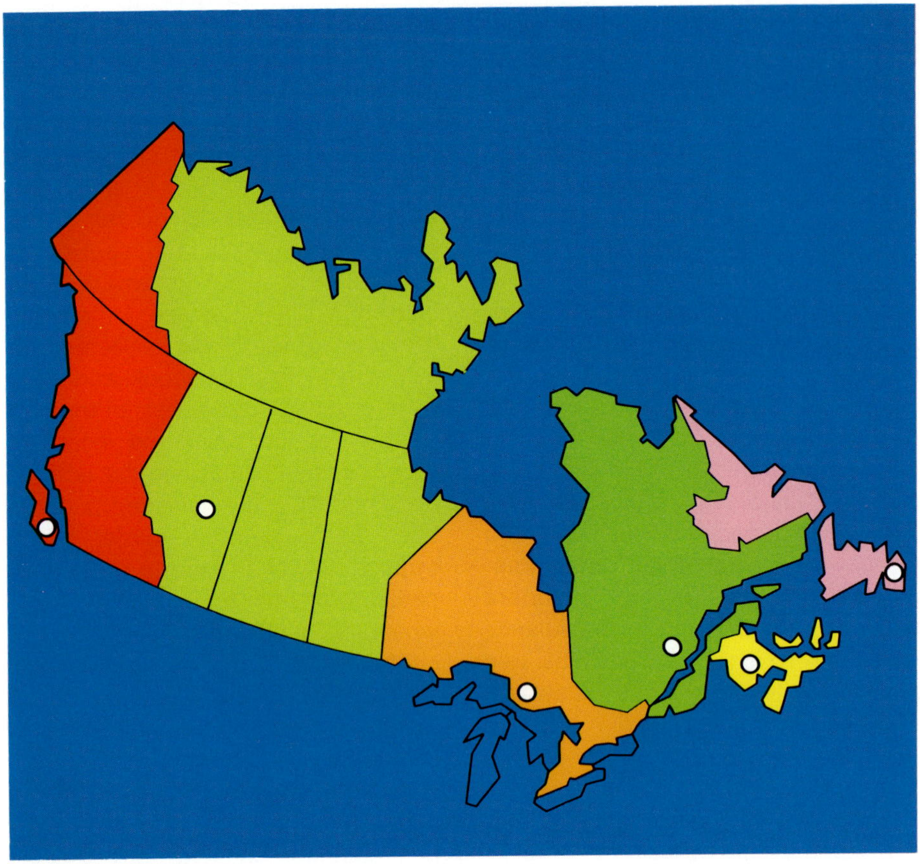

Forestry Centres

Pacific Forestry Centre
506 West Burnside Road, Victoria, British Columbia V8Z 1M5

Northern Forestry Centre
5320-122nd Street, Edmonton, Alberta T6H 3S5

Great Lakes Forestry Centre
Box 490, Sault Ste. Marie, Ontario P6A 5M7

Laurentian Forestry Centre
Box 3800, Ste. Foy, Quebec G1V 4C7

Forestry Canada — Maritimes
Box 4000, Fredericton, New Brunswick E3B 5P7

Newfoundland Forestry Centre
Box 6028, Building 304, St. John's, Newfoundland A1C 5X8

with the tree's environment either above or below ground. With a few exceptions, the following account will deal with problems created by insects. However, some common agents of injury other than insects will be treated superficially to clarify doubts concerning cause. Because insect problems involving seedlings in a nursery are quite distinctive and generally concern only a limited number of people, they are excluded from this publication.

There is no part of a tree that is not subject to attack by some insect; the degree of injury inflicted on the tree, however, is dependent on the number of insects, type of feeding, time of year, point of attack and how vital to survival is the part attacked. Usually, the earlier the damage can be detected, the greater is the chance of determining the causal agent, since secondary insects often move in to feed on dead or dying parts. In all cases of foliage discolouration and wilting, the insect causing the damage, or clues to its identity, will be found at the junction of living and dead tissues when examination is carried out early enough.

Control

The Forest Insect and Disease Survey is regularly asked to recommend insecticides for the control of insects on trees. In contrast to the wide use of chemical control measures that seemed a few years ago to offer immediate freedom from pest species, a more cautions approach is now being taken. Moreover, some chemical pesticides are no longer available and more restrictions are regularly being imposed on others as undesirable side effects are discovered. Currently, other methods are being sought that place greater emphasis on biological control or are more selective and less deleterious to the environment. Because of these changes, no specific control measures are given in this handbook. On the basis of past experience however, the necessity for control and the stages in the pest's life cycle most susceptible to treatment are pointed out. Since the type of feeding (chewing or sucking) and habit of the insect (open feeding or mining) will also influence the selection of control measures, suggestions are made regarding the required mode of action (contact, stomach, systemic or fumigant) of the pesticide. If large-scale chemical control measures are necessary, the advice of specialists should be sought regarding the currently acceptable pesticide. The registered uses, the concentration to be used and the precautions to be taken are listed on the labels of pesticide containers.

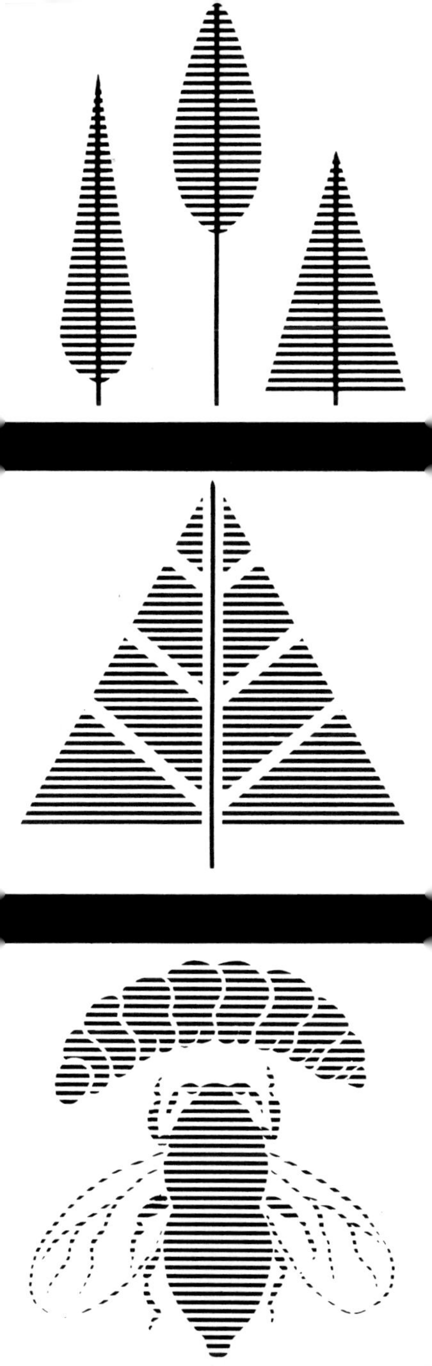

Spruces, fir and hemlock

White spruce

Black spruce

Norway spruce

Red spruce

Colorado or blue spruce

White, black, red, Norway and Colorado spruce needles

Balsam fir

Eastern hemlock

Spruce, fir and to a lesser degree, hemlock are common components of forests in eastern Canada. Of the three native species, white spruce and black spruce are transcontinental in range whereas red spruce has a much more limited distribution, being found commonly in the Maritime Provinces and in a wide band on either side of the St. Lawrence River in Quebec, with scattered pockets in eastern Ontario. The two introduced species, Norway spruce and Colorado spruce, are widely planted ornamentals; blue forms of the latter, particularly the form known as Koster, are held in high regard. Balsam fir occurs from the Rocky Mountains to the Atlantic, whereas eastern hemlock has a more restricted distribution along the southern boundary of northwestern Ontario and eastward to the Maritime Provinces. Additional information on all of the native species will be found in the Canadian Forestry Service publication *Native Trees of Canada,* available in many bookstores as well as from Supply and Services Canada, Ottawa.

Parts of a tree

Insects generally restrict their feeding to specific parts of a tree. Therefore, if the parts can be identified it is possible to separate the large number of insect species into manageable groups. Some parts of a tree grade imperceptibly into each other and are best demarcated by an understanding of the seasonal growth of a tree. In winter both terminal and lateral buds are found on all species. All of the spruces and balsam fir have a relatively large terminal bud usually situated between two laterals. On vigorous growth there will be, in addition, a number of lateral buds scattered along the twig. Buds of hemlock, both the terminals, which are single, and the scattered laterals, are quite inconspicuous. In the spring the buds initially swell and then elongate, revealing the closely packed tiny needles of the current year's shoots. Soon the needle-like leaves separate and the silhouette typical of the spruce, fir or hemlock is revealed. Until the tissues harden, the new shoots have a characteristic droop. By midsummer the buds that will open the following year are present on the shoots. Male (staminate) flowers, which release their yellow pollen as the shoots are elongating, are in the needle axils along the previous year's growth, in dense clusters on balsam fir, but more scattered on the spruces and hemlock. Some of the terminal buds on spruce and hemlock develop into cones that mature within the year. On the other hand, the upright balsam fir cones are borne along year-old twigs. Cones and needles of all the species are illustrated on pages 16 to 18. In this handbook "shoot" is defined as

Parts of a tree

linear growth of the current year and "twig" as that of the previous year. A "branch" is any portion of the crown that is older than 2 years, excluding the main stem, the apical portion of which is called the "leader". The parts of a tree illustrated are those of the common white spruce.

Types of insects

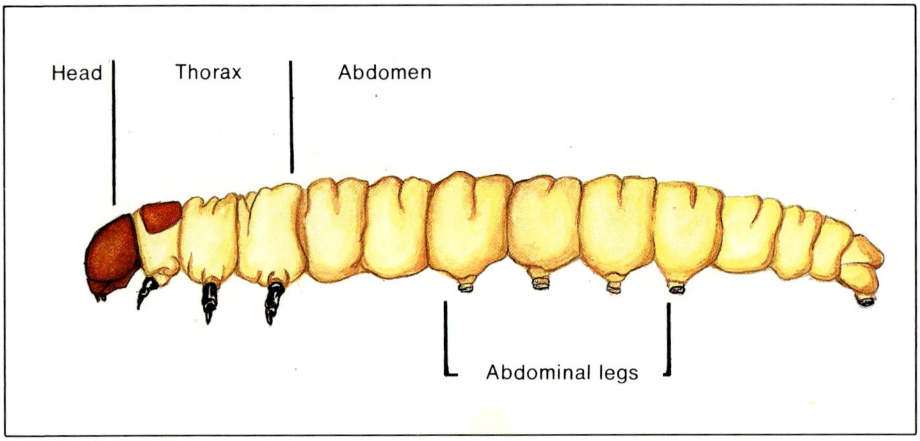

Will be a moth

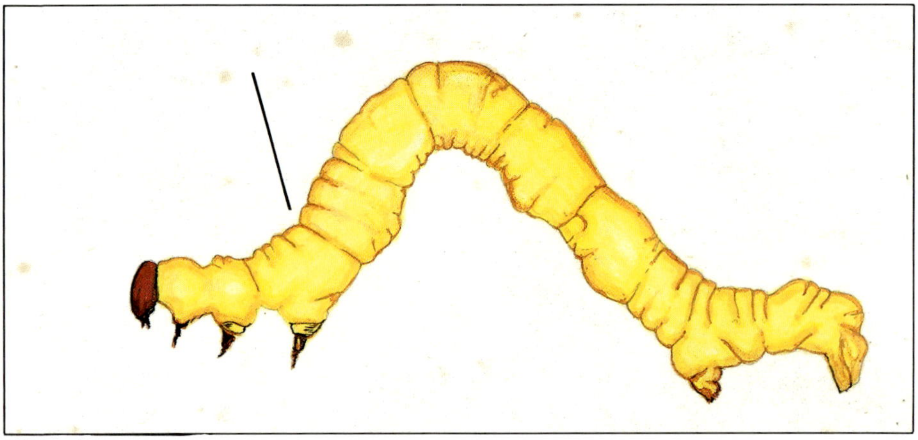

Will be a looper moth

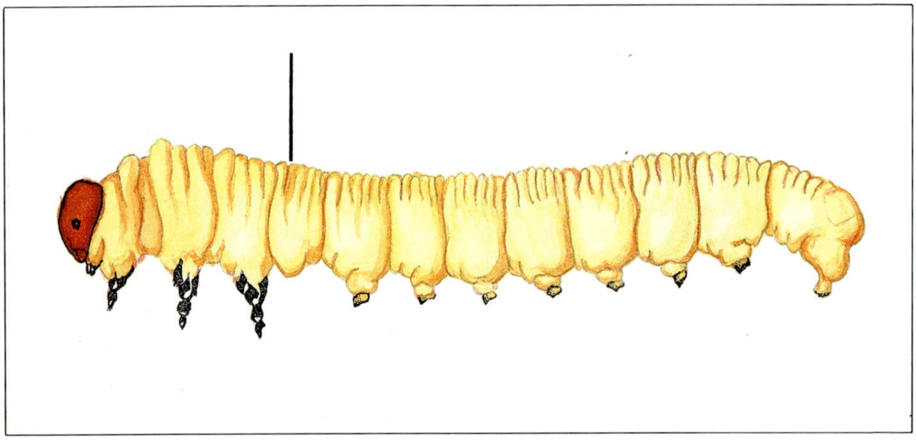

Will be a sawfly

Insects develop from egg to adult by either complete or simple metamorphosis. In complete metamorphosis, the insect develops through four stages – egg, larva, pupa and adult – each different from the other. This is how most of our destructive forest pests develop. The larva, also referred to as "worm", "grub" or "maggot", is the principal feeding stage. It sheds its skin periodically from the time it emerges from the egg as a tiny individual until it achieves full growth. The pupa is an inactive, transformation stage between the feeding larva and the reproducing adult. Larvae, particularly those destined to become moths, vary greatly in appearance, shape and size, and may be hairy or naked. The majority of our forest insects can be grouped according to the general structure of the larva as indicated in the simplified outline drawings □.

In simple metamorphosis the insect develops through three stages: egg, nymph and adult. The nymphal stage□ is the active feeding stage and the nymph, which usually resembles the adult, sheds its skin periodically as it grows. A forest insect in this group is the pine spittlebug (see page 104).

Will be a beetle

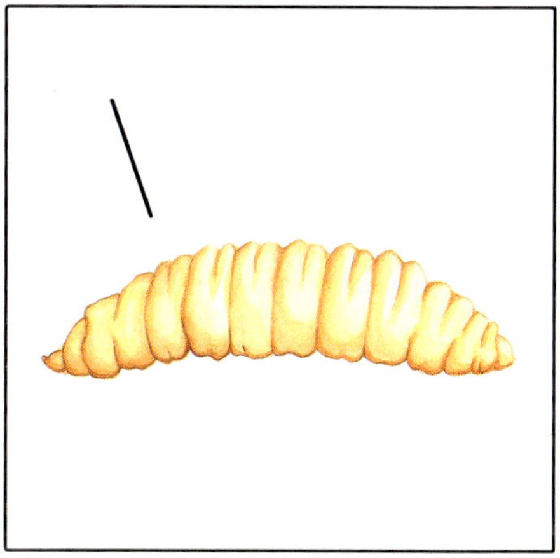
Will be a two-winged fly (head doesn't show)

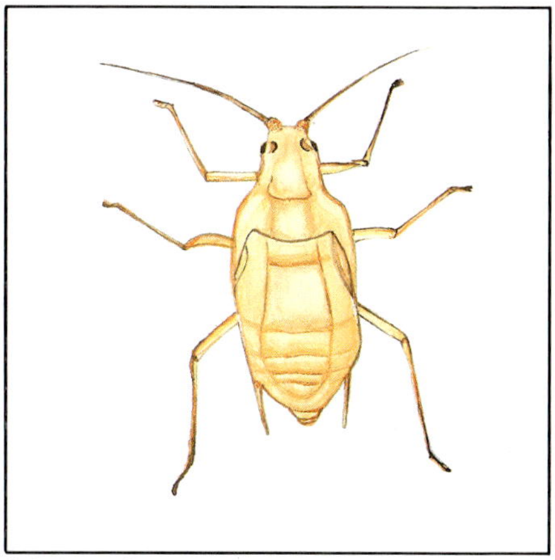
Aphid nymph

Ant

Lady beetle

Click beetle

Not all insects found on trees should be regarded as pests. This is especially true of adults found during the day, when beneficial or innocuous species are often present. Some common adult insects found on trees are: ants □, often associated with aphids, which they tend; lady beetles □, which are beneficial since they feed on harmful insects; and click beetles □, whose larvae live in soil. In addition, March flies and May flies are sometimes briefly abundant on trees at the time of mating flights. Adults of most harmful species feed and lay their eggs mainly at night and are not often seen.

Galls caused by insects

Needle galls

Needle and shoot gall

Shoot and twig gall

Insects cause damage not only by consuming tissue or sap but also by stimulating abnormal development, resulting in galls, during cell division and growth. Insect-induced galls range from a simple kink in a leaf to complex structures bearing little resemblance to what should have been. Because of the wide variety of galls produced, it is probable that more than one mechanism is involved, but the actual inciter of gall formation generally remains a mystery. Three examples of galls on spruce and fir are shown. Information on a needle gall is given on page 93, needle and shoot galls on pages 43 to 48, and twig galls on pages 48 and 105.

The following pages show how to identify insects on or injury to various parts of a tree.

Bud or shoot

*Dead buds are sometimes found without any evidence of insect or disease.

**This may also be caused by unsatisfactory site or weather conditions.

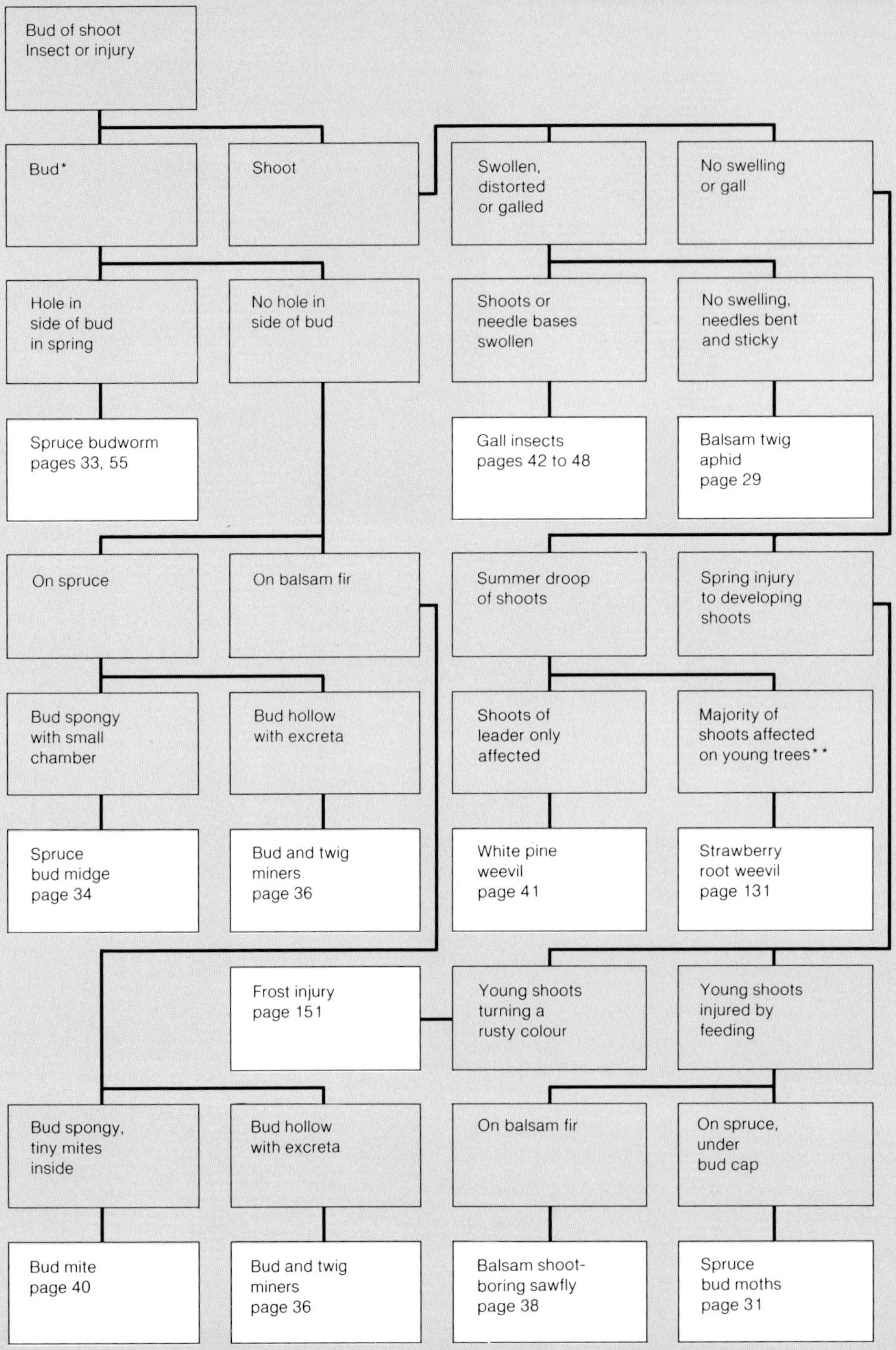

Balsam twig aphid

Damaged shoot

The balsam twig aphid, *Mindarus abietinus* Koch, is found wherever balsam fir grows and the same or a closely related species is found on white spruce. These aphids are often very abundant on fir, but infestations are usually of short duration. However, shoot and needle injury, which gives the foliage a ruffled appearance, lowers the quality of fir and spruce cut as Christmas trees.

The overwintered eggs hatch in late April or early May and first generation nymphs feed on needles near the buds. Nymphs of the second and third generations feed on needles of the developing shoots □ in May and June, excreting masses of waxy white "wool" □ and large quantities of a sticky substance known as "honeydew". The 3-mm-long winged adults □ that develop fly to other trees, where each produces about 10 living young. These fourth-generation nymphs feed lightly for a short period, producing wingless adults that lay one or two black eggs around the buds.

Predaceous larvae of lady beetles □ and flower flies generally provide

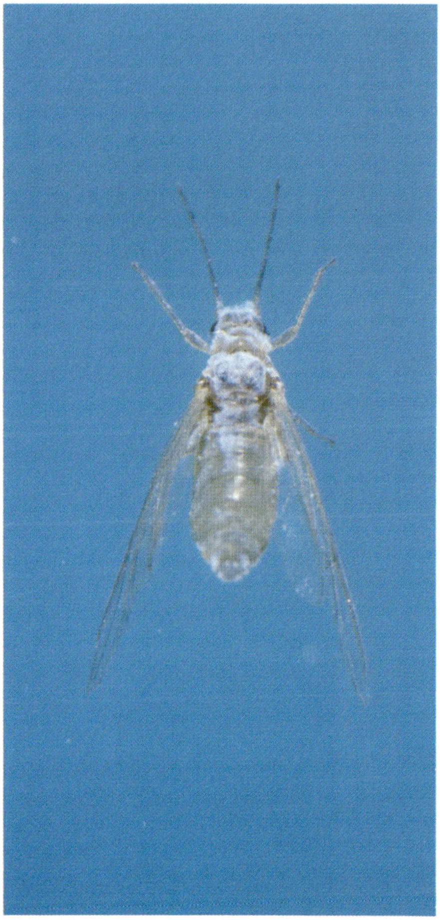

Adult

Early feeding

Predator feeding on aphids

adequate control. However, where young balsam fir or spruce are to be harvested as Christmas trees, chemical control of the aphid may be required in some years. For currently acceptable control measures and proper timing of application, specialists should be consulted.

Spruce bud moths

Early damage

Feeding damage complete

The spruce bud moths of the genus *Zeiraphera* are transcontinental in range. They are found frequently on white spruce but only occasionally on other spruces and firs. Although severe feeding damage has seldom been extensive in forests, new shoots on open-grown white spruce trees in plantations or on ornamentals are frequently disfigured.

Of the three closely related species involved, *Zeiraphera canadensis* Mutuura & Freeman is probably the most common. It overwinters in the egg stage on the tree and in May the larva emerges and starts feeding in the developing bud. The larva characteristically secures the bud cap to the growing shoot □ with silk and the cap remains there long after other bud caps are shed. When populations are high, it is not unusual to find several larvae feeding among the needles of a single shoot. The full-grown larva □ is about 10 mm long and cream to yellow, although the head and thoracic legs are dark brown on younger larvae.

canadensis larva

fortunana larva

unfortunana larva

Immature spruce budworm (*Choristoneura fumiferana*)

When feeding is completed □, in June or early July, the larva drops to the ground, where it is transformed into a pupa in the litter. Moths emerge from the pupae from mid-July to mid-August and the eggs are laid singly or in groups of two to seven at the base of the current year's shoots.

The larvae of the two related species *Zeiraphera fortunana* (Kearfott) □ and *Zeiraphera unfortunana* Powell □ feed at the same time of year and in a manner similar to that of *Zeiraphera canadensis* □. In addition young spruce budworm larvae □ treated in detail on pages 54 to 59 may be found in developing shoots.

Although many parasites and predators feed on the larvae of the spruce bud moths, chemical control may be required to preserve the aesthetic value of trees. A contact insecticide with some fumigant action, if applied when the buds begin to elongate, will reduce the damage to a minimum.

Spruce bud midge

Injured bud

The spruce bud midge, *Rhabdophaga swainei* Felt, feeds on all species of spruce thoughout its transcontinental range. Red spruce within its range is the preferred host; elsewhere it is most commonly found on black spruce. The central bud at the tips of twigs on open-growing trees is most often killed by the midge □. This usually creates a pruned effect and sometimes results in multiple leaders in young trees.

The orange-red female midge □ lays her eggs in the developing shoots in May or June. On hatching, the young larva bores into the tip of a new shoot and later enters the newly formed bud, where it feeds until winter. In the spring the pink, legless larva, about 2 mm long, can be found in a small cell in the centre of a rosette-like bud between two developing shoots □. The red and brown midge pupa formed in the larval cell pushes its way towards the bud tip until it protrudes, thereby allowing the adult to escape from the pupal case.

A number of species of tiny wasp-like parasites kill the midge larvae and play an important role in controlling this pest. Hence, a tiny whitish larva or black pupa of the parasite is commonly found in place of the host larva in the damaged bud. Serious, lasting injury to spruce caused by the spruce bud midge has not been recorded and the aesthetic value of ornamental trees is not decreased to any extent.

Larva in bud

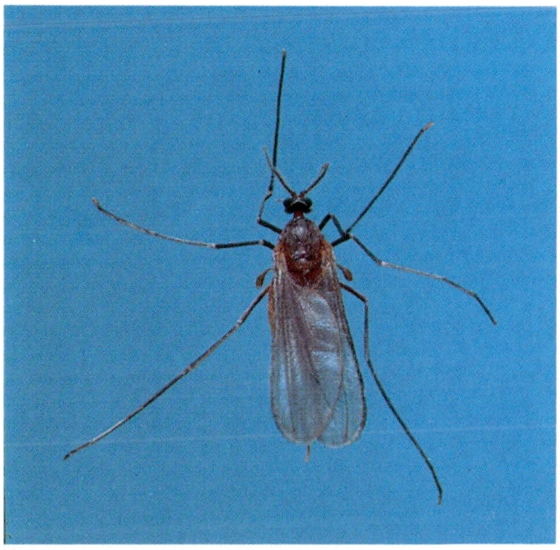

Adult

Bud and twig miners

Damaged black spruce

Three closely related bud and twig miners occur on spruce and balsam fir. The species *Argyresthia mariana* Freeman on black spruce □ and *Argyresthia abies* Freeman on balsam fir □ are known only in northern Ontario. The third species, *Argyresthia picea* Freeman, has been found on white spruce in the Yukon Territory, Ontario and southern Quebec. No serious injury to any of the host trees has been recorded.

The larva overwinters in the bud. In the spring, when it is full-grown □, it is about 5 mm long and cream coloured with blackish head and legs. It changes to a pupa in the mined tips from mid-May to late June and the tiny adult emerges shortly afterward from a silk-covered hole □ in or below the base of the dead bud. The two species on spruce complete one generation each year, but the species on balsam fir appears to require 2 years for each generation.

No control has been required for these insects.

Damaged balsam fir

picea larva

Silk-covered adult exit

Balsam shootboring sawfly

Damage

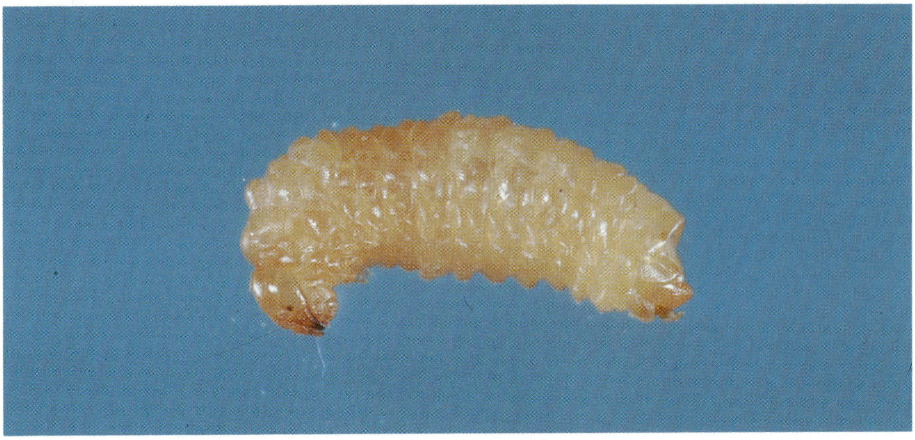

Larva

The balsam shootboring sawfly, *Pleroneura brunneicornis* (=*borealis*) Rohwer, has been reported from Alberta to Nova Scotia and south of the Great Lakes. The larvae bore into and kill the new shoots of balsam fir □, causing damage similar to that caused by late spring frost. Although damage is often conspicuous, no associated tree mortality has been reported.

The adults □ emerge in the spring and eggs are inserted singly into the tightly packed needle clusters shortly after the bud scales have dropped. Feeding larvae have been found burrowing into shoots from about mid-May to early July. The whitish larva, about 6 mm long when full grown □, drops to the ground and spins a cocoon, inside which it overwinters as either a larva or a pupa. In Ontario it appears that this insect spends a full year in the cocoon in the ground, since feeding larvae are abundant only every second year in any one locality.

Control measures have not been required for the balsam shootboring

Adult

sawfly. However, in the event that they become necessary, a systemic type of insecticide would perhaps be most appropriate, since the larvae are concealed feeders.

Bud mite

Damaged bud

The bud mite, *Trisetacus grosmanni* Keifer, is a microscopic, worm-like creature with two pairs of legs that is commonly found in buds of balsam fir and occasionally, white spruce in Ontario. Since it is also found on Sitka spruce on the west coast, it will probably be found on other spruces. It is also associated with bud proliferations on red pine twigs.

Little is known of the life history of this mite. Eggs, larvae and adults are found inside buds ☐ that have failed to undergo any development in the spring ☐. The condition of these buds suggests that they were probably killed in the fall of the previous year. The amount of damage done by this mite to date has not been sufficient to justify the use of chemical control measures.

Mites in bud

White pine weevil

The white pine weevil, *Pissodes strobi* (Peck), and its varieties, are serious pests particularly of pine, but also of spruce, fir and hemlock, throughout Canada and much of the United States. In eastern Canada, Norway spruce is often severely injured. This insect is particularly damaging, since the larvae kill the leading portion of the stem, thus adversely affecting the form of trees attacked.

The adult weevils hibernate in the litter under infested trees and emerge in early spring. Initially they feed on the upper part of the leader, and after mating the female lays eggs there in feeding punctures. The pale, legless larvae, about 7 mm long when full grown, tunnel down the leader causing the top whorl of shoots to wither and die, and thus to form a typical shepherd's crook □. The larvae change to pupae under the bark at the feeding site in the wood or pith produced the previous year. The adults emerge from the infested stem in August and September and after some feeding go into hibernation. A more extensive treatment of the white pine weevil may be found in the first handbook of this series, *Insects of Eastern Pines.*

This weevil may be controlled by pruning and burning infested leaders in early July. Also effective are stomach poisons with good residual properties applied to the leaders in early spring, when the adults are first observed.

Damaged Norway spruce

Gall insects

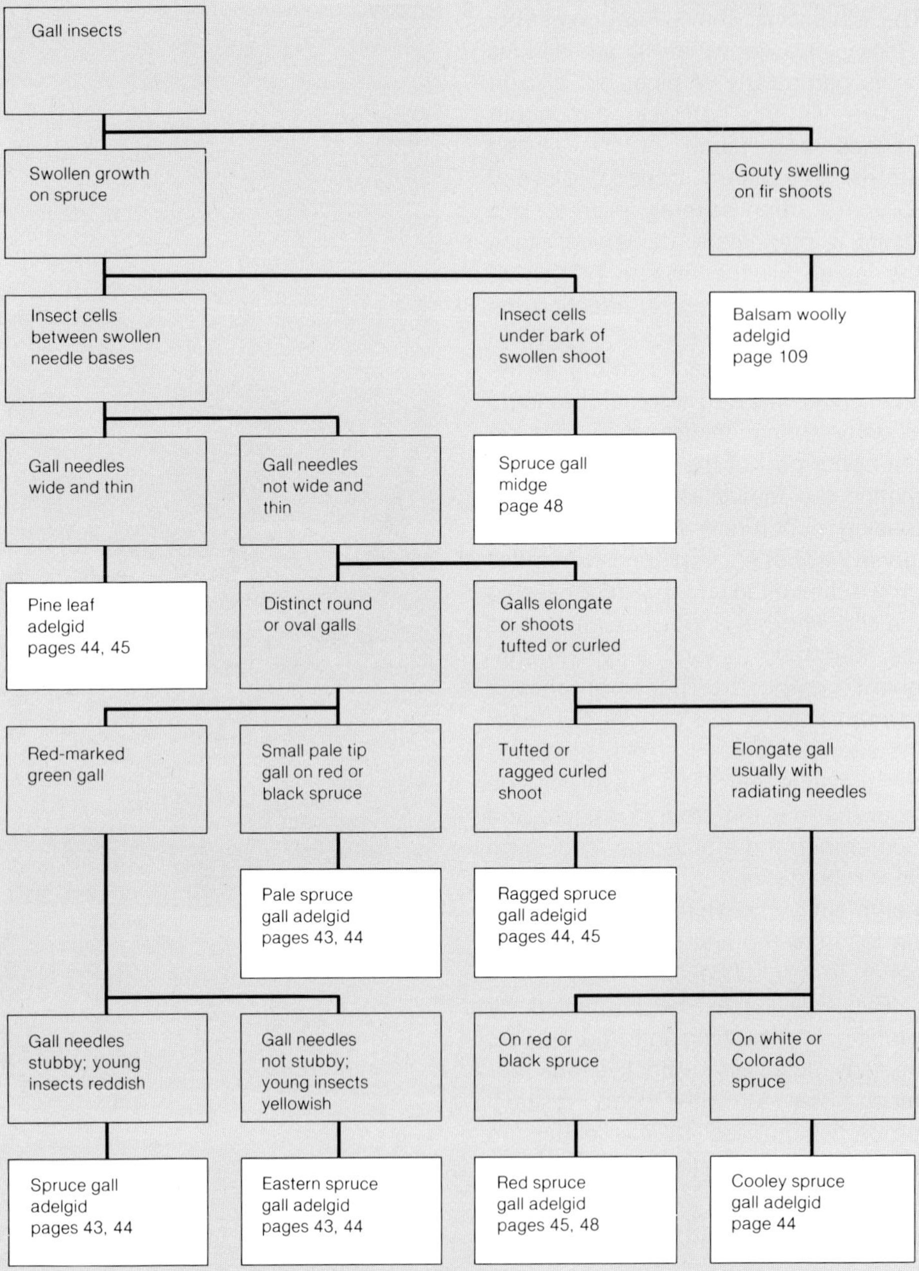

Eastern spruce gall adelgid

Spruce gall adelgid

Pale spruce gall adelgid

With the exception of the spruce gall midge, the galls dealt with in this section are caused by aphid-like insects of the family Phylloxeridae (Adelgidae) commonly known as the spruce gall adelgids. They have complex life cycles, some species feeding exclusively on spruce while others feed on spruce and one other conifer. However, galls characteristic of each species are formed only on spruce. Six generations are usually required to complete the 2-year cycle, and in the case of species with alternate hosts, winged adults about 2 mm long are formed only in the generations that move from one host to the other. A simplified life cycle of a species that alternates between spruce and larch is illustrated on pages 46 and 47.

The **eastern spruce gall adelgid** □, *Adelges abietis* (Linnaeus), an introduced species, feeds only on spruce and is currently found from Ontario eastward and in the adjacent areas of the United States. The **spruce gall adelgid** □, *Adelges lariciatus* (Patch), occurs in alternate years on spruce and larch from Alberta to the Maritimes and in adjacent areas in the United States. The **pale spruce gall adelgid** □, *Adelges strobilobius* (Kaltenbach), also occurs from Alberta to the Maritimes and in adjacent areas of the United States, usually alternating between black or red spruce and larch. The **Cooley spruce gall adelgid** □, *Adelges cooleyi* (Gillette), is mainly a western species that usually alternates between white spruce and Douglas-fir. In eastern Canada it is rare and in Ontario the galls are usually found on Colorado spruce.

Cooley spruce gall adelgid

Members of the insect genus *Pineus* also produce galls on spruce. The **pine leaf adelgid** □, *Pineus pinifoliae* (Fitch), occurs from Alberta to the Maritimes and in the eastern United States. It has caused serious injury to the alternate host, eastern white pine. The **ragged spruce gall adelgid** □, *Pineus similis* (Gillette), occurs across the continent. It has no alternate host and repeated attacks by the insect often result in dead twigs (see page 105) and a generally ragged appearance of the foliage. The **red spruce**

Pine leaf adelgid

Ragged spruce gall adelgid

Red spruce gall adelgid

Simplified aphid life cycle on alternate hosts

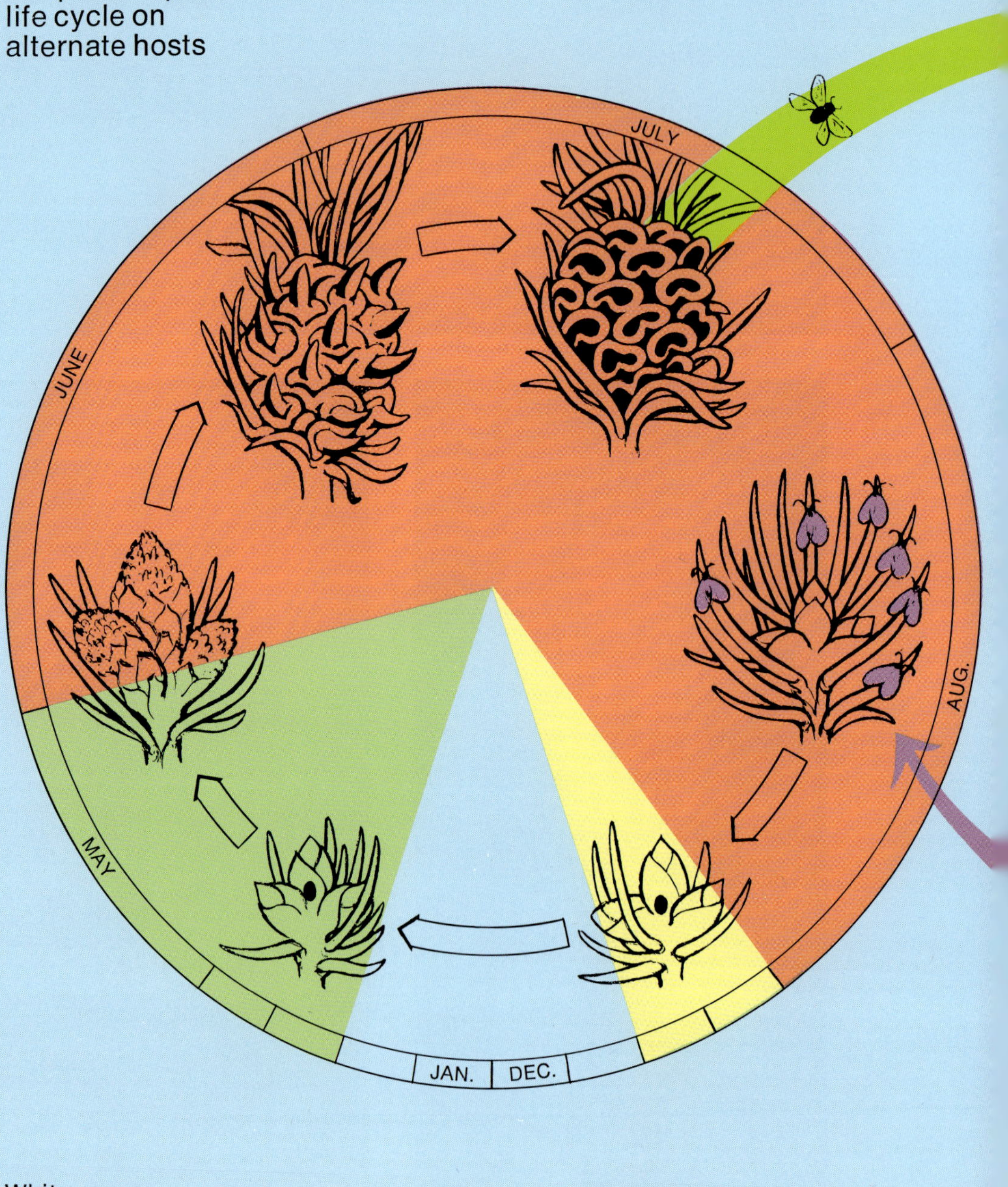

White spruce

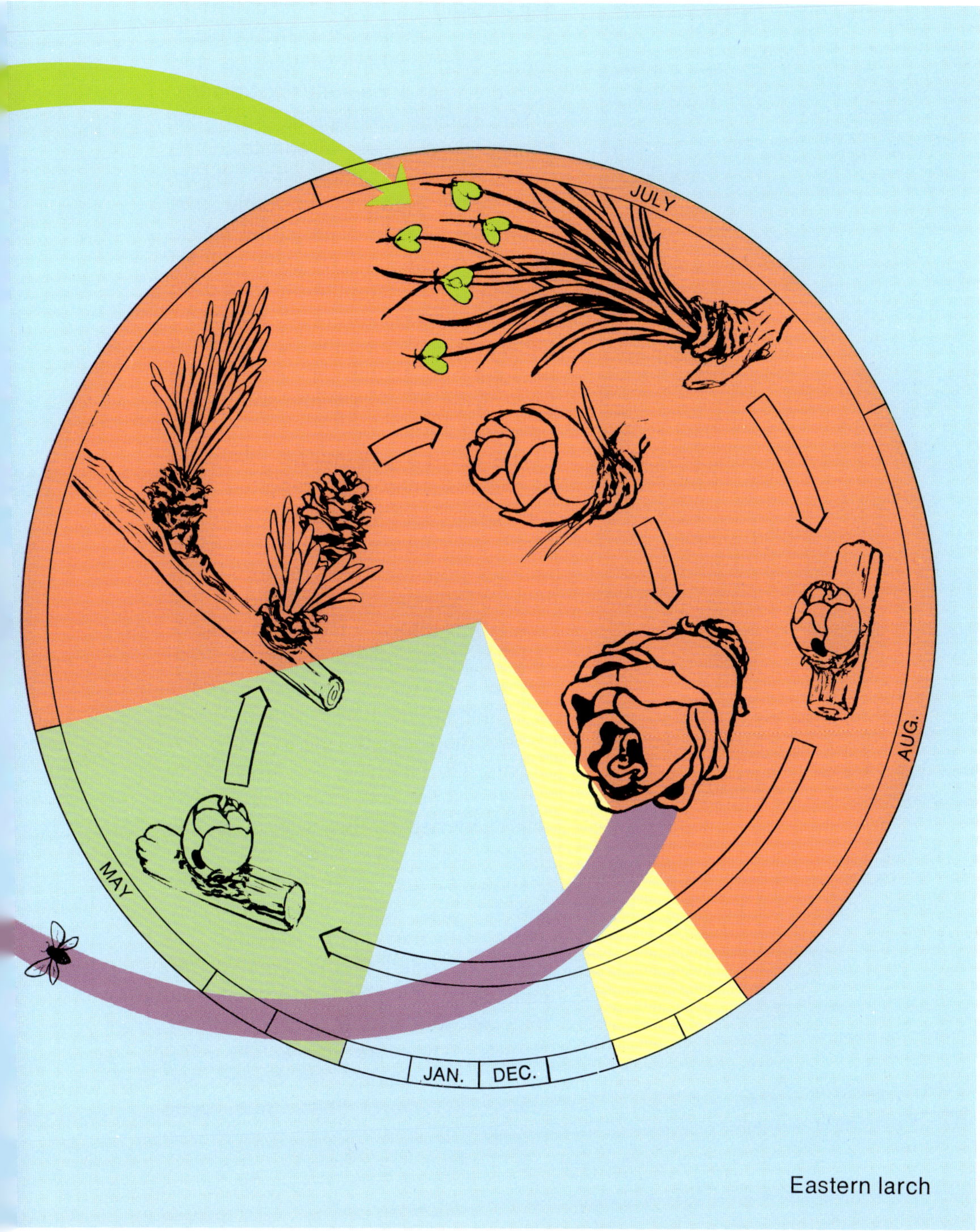

Eastern larch

Spruce gall midge injury to shoot

gall adelgid □, *Pineus floccus* Patch, alternates between either red or black spruce and eastern white pine and outbreaks are apparently uncommon.

The spruce gall aphids have not been a serious pest of spruce in the forest. However, they often render ornamental plantings unsightly and retard the growth of potential shade trees. Control can be effected by pruning and burning the green, closed galls in spring and early summer. Otherwise an insecticide with systemic or fumigant action, applied as the buds begin to swell in the spring, will provide control.

The **spruce gall midge**, *Mayetiola piceae* Felt, has been reported from the Yukon Territory and Alberta to the Maritimes. Severe injury to spruce is unusual and has been recorded only from a number of locations in the Yukon Territory. The two-winged, reddish-brown midge emerges from the galls formed in the previous year's shoot (see page 105) in late May, and after mating, lays her eggs on the new shoots. On hatching, the larvae bore into the shoot and form cells □. The gall is formed by the swelling of the tissue about the larval cells. The orange larvae overwinter in the shoots, and when full grown, are about 1.5 mm long. Although the pupa is formed in the larval cell in early spring, it often protrudes from the cell before the adult emerges. Usually tiny parasitic wasps keep midge populations in check and chemical control has not been required.

Undamaged white spruce buds

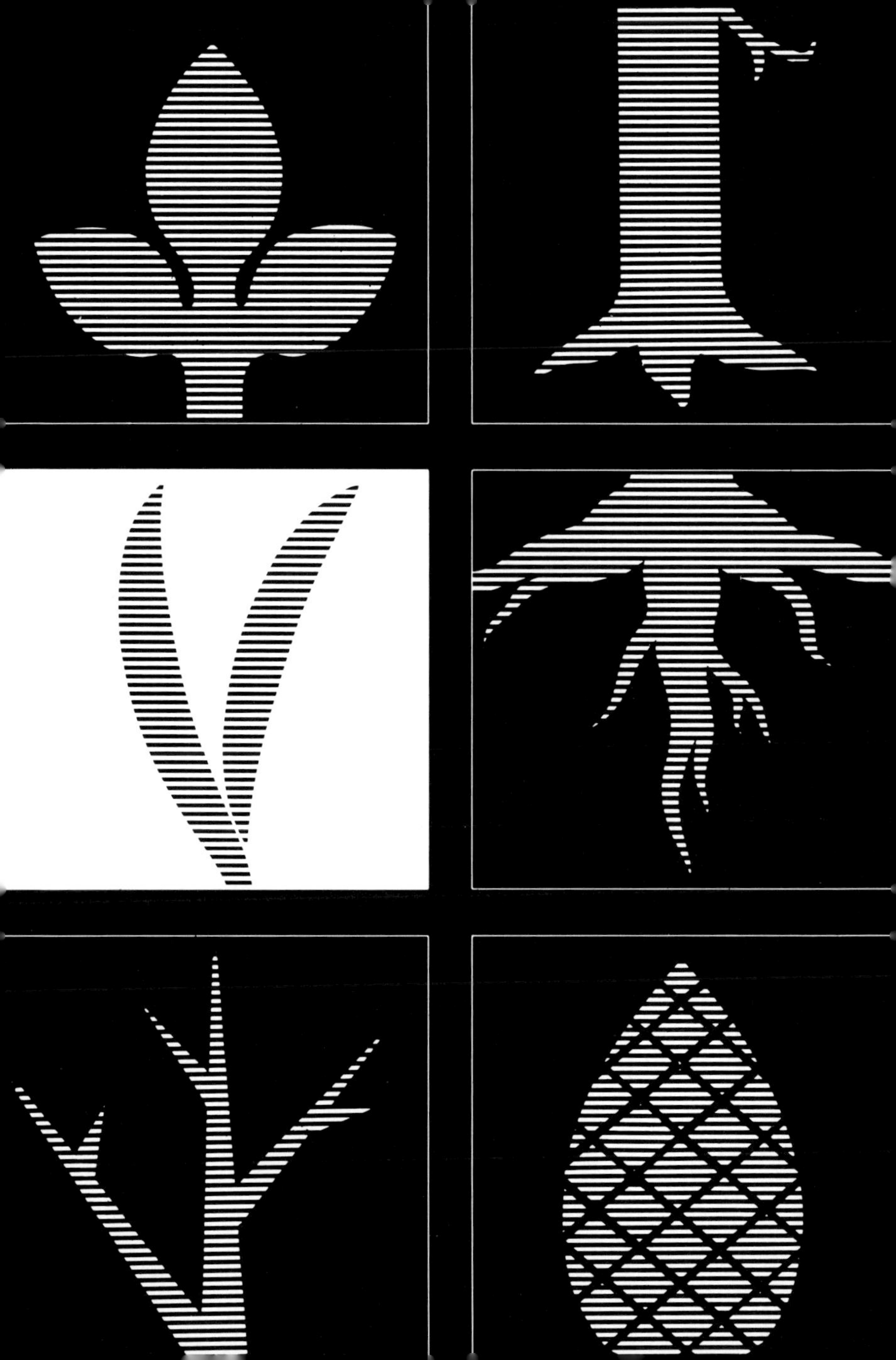

Needle

*Solitary larvae without obvious feeding injury are included here but not needle miners.

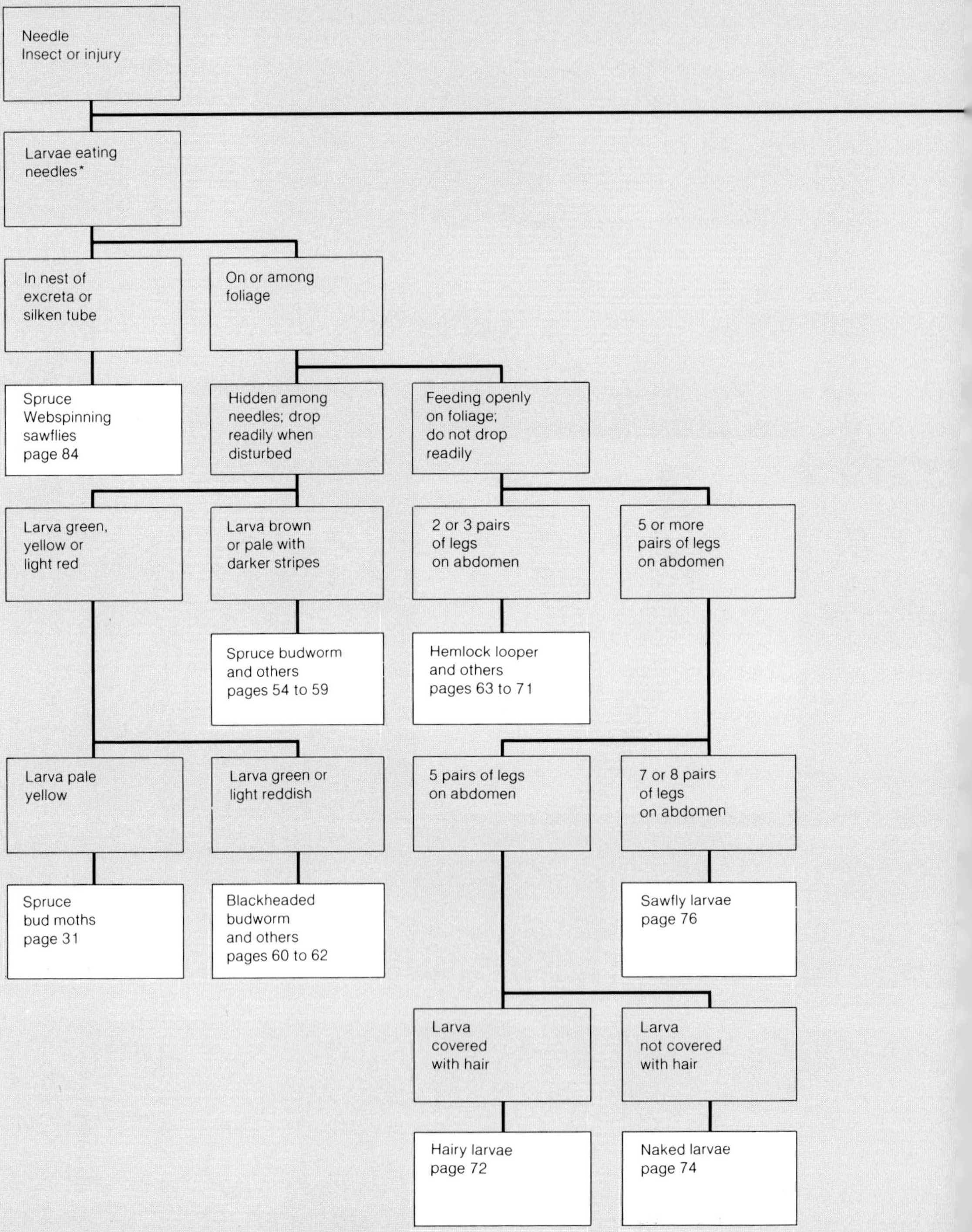

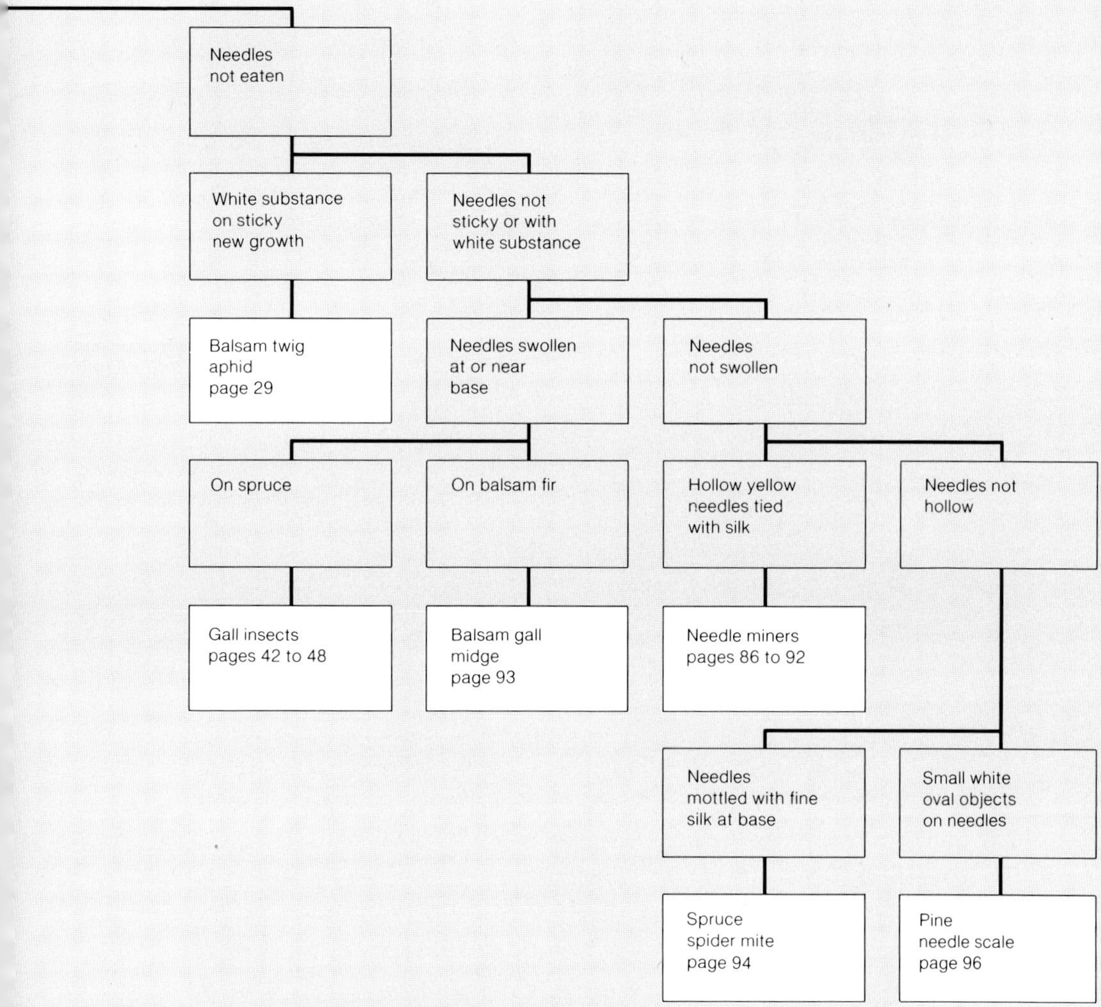
Needles not eaten
White substance on sticky new growth
Needles not sticky or with white substance
Balsam twig aphid page 29
Needles swollen at or near base
Needles not swollen
On spruce
On balsam fir
Hollow yellow needles tied with silk
Needles not hollow
Gall insects pages 42 to 48
Balsam gall midge page 93
Needle miners pages 86 to 92
Needles mottled with fine silk at base
Small white oval objects on needles
Spruce spider mite page 94
Pine needle scale page 96

Spruce budworm and others

Spruce budworm

Spruce coneworm

Spruce bud moth larva

Spruce budworm

• Larva brown or with rows of small pale spots on brown body ....spruce budworm.
• Larva striped rusty brown .... spruce coneworm, pages 136, 137
• Larva pale yellow with darker stripes .... spruce bud moth (*Zeiraphera unfortunana*), pages 32, 33

The spruce budworm, *Choristoneura fumiferana* (Clemens), is the most destructive forest insect in North America. In eastern Canada, massive epidemics of this pest occur periodically in spruce-fir forests, resulting in losses of millions of cubic metres of spruce and fir. Mature stands of balsam fir are particularly susceptible and tree mortality occurs after several years of continued heavy feeding. In eastern Canada the spruce budworm feeds primarily on balsam fir and red and white spruces, but damage to other conifers occurs when these are growing in mixture with the favoured tree species.

The budworm overwinters on the tree as a very tiny larva in various hibernation sites: in male flower cups □, under bud and bark scales, amongst lichens or in bark crevices. Shortly before the buds begin to expand in the spring, the larvae emerge from their winter quarters and mine needles produced the previous year (see page 87), unopened buds or, when available, male flowers (see page 144). Later they all feed in the expanding buds □ and, as the new shoots grow, they spin fine silk among the needles and between shoots. When the new needles have been eaten they move back to feed on the

Overwintering larva

Bud-mining larva

Severe defoliation

older needles □. Heavy feeding will cause trees to take on in midsummer a scorched appearance that can be readily seen from aircraft □. When the larvae are fully grown □ and about 22 mm long, they change to pupae □ in the feeding sites or on lower branches. The pupae become moths □ from late June to early August and after mating the females deposit eggs in masses of 15 to 50 □ on the underside of the needles. The eggs hatch in about 10 days and the tiny larvae soon seek hibernation sites in which to spin their silken shelters and pass the winter.

During epidemics of the spruce budworm, distant forests may be invaded as a result of long-range moth flights. In addition, when young larvae hatch in summer or when they come out of hibernation in the spring, they tend to spin down on long silken threads and, on being picked up by convective air currents, may be widely dispersed.

Pupa

Adult

Egg cluster

Numerous kinds of parasites □ and predators feed on the spruce budworm during all of its life stages, and disease organisms □ take an additional toll. However, despite these natural control factors, large-scale chemical control operations in various parts of eastern Canada have been necessary to protect valuable spruce and fir stands. A great deal of research is also being conducted to provide forest managers with alternatives to chemical insecticides. These include enhancement of the killing

Larva killed by parasites

Larva killed by fungus disease

capability of naturally occurring disease organisms such as fungi, bacteria and viruses, the use of chemical agents to produce sterility, the use of sex attractants for early detection of population increases, and the prevention of reproduction by the use of juvenile hormones, to mention a few. While this research will undoubtedly supply us with effective new approaches to budworm control, it is likely that, owing to the enormity of the problem, chemical insecticides will continue to play a major role in spruce budworm management in the foreseeable future.

Large-scale chemical control operations carried out by forest managers and involving millions of hectares and fleets of aircraft in a single year are beyond the scope of this handbook. On the other hand, where a limited number of valuable shade or ornamental trees are involved, damage can be kept to a minimum and the insect more readily killed if the insecticide is applied when the larvae are

Aerial view of heavy defoliation

small. Therefore, a contact or stomach insecticide should be applied soon after the buds have burst.

Blackheaded budworm and others

Blackheaded budworm

Spruce needleworm

Redstriped needleworm

• Larva with green body and red to near-black head, up to 15 mm long .... blackheaded budworm.
• Larva with green body and pale to light-brown head, up to 22 mm long .... spruce needle worms.
• Larva pale reddish, indistinctly striped, up to 10 mm long .... redstriped needleworm.

The blackheaded budworm, *Acleris variana* (Fernald), has been reported from Alberta to Newfoundland; however, epidemics of the pest have occurred only in eastern Quebec, the Maritimes, Newfoundland and the northeastern United States. The preferred host tree is balsam fir, but the larvae also feed on spruce and hemlock.

The insect overwinters in the egg stage on needles of the host trees and hatching occurs from mid-May to early June. The larvae feed in the elongating buds and later, on needles of the new shoots. When full grown, they are about 15 mm long and the head colour varies from reddish to dark brown, sometimes with a mottled pattern ☐. When feeding is complete the larvae change to pupae, usually in the final feeding site ☐ page 62. The adult moths emerge from the pupal cases in July and August and the eggs are laid singly, usually on the underside of the needle. The colour pattern of the wings of the black-headed budworm moth is unusually variable.

Generally parasites appear to play a major role in population control. Unfavourable weather during the late larval stages often causes additional mortality and large-scale chemical control measures have not been necessary to date. Larvae in home plantings can be prevented from causing damage by application of a contact or stomach type of insecticide shortly after the buds have burst.

The **spruce needle worms** treated here are commonly found in small numbers on spruce, fir or hemlock throughout most of Canada and the northeastern United States. The best known is *Archips packardiana* Fernald. It overwinters as a tiny larva in a needle mine. Needle mining is resumed in the spring and later the larvae move to feed on the new foliage, where they spin considerable webbing. When full grown, the larva ☐ is about 20 mm long and has a pale green head sometimes patterned with brown, a pale body and pale thoracic legs. It changes to a pupa usually in the webbed needles. The adults emerge from summer to early fall.

The closely related but much less common species, *Archips strian* (Fernald), probably has a habit similar to that of the foregoing species. The larvae, however, have dark thoracic legs and small dark areas around the base of the hairs on the thorax.

Another needle worm, *Clepsis persicana* (Fitch), which feeds on a wide variety of coniferous and deciduous trees, is occasionally found on spruce and fir. It is a slender very active larva, with yellow head and indistinct, wide greyish stripes on the body. In Ontario the larvae have been collected in late May and June.

Control measures have not been necessary for the spruce needle worms. If they should become abundant on specimen trees, a contact or stomach type of insecticide recom-

Blackheaded budworm pupa

mended for defoliators should be applied when the new shoots begin to develop.

The **redstriped needleworm,** *Griselda radicana* Heinrich, occurs commonly in the spruce-fir forests of North America, but there are no records of serious outbreaks. White spruce is the preferred host tree.

The eggs overwinter on the needles and the larvae feed on the new foliage from late May to late July. The full-grown larvae □ are about 9 mm long. They drop to the ground, where they change to pupae in silken cocoons in the litter. The adults emerge in late summer and fall and the females lay their eggs singly at the base of the needles. No control measures have been necessary.

Hemlock looper and others

More than a dozen kinds of looper feed on the spruces, fir and hemlock in eastern Canada. By using the season of occurrence and illustrations shown on succeeding pages, most of them can be identified and dismissed because they are usually of little consequence. On the other hand, the hemlock looper is considered a pest of primary importance and is treated in greater detail.

Although the hemlock looper, *Lambdina fiscellaria fiscellaria* (Guenée), occurs from Alberta eastward, severe, prolonged outbreaks have been recorded only in the east from southern Ontario to Newfoundland and in the northeastern United States. It is the main pest of balsam fir in Newfoundland, where, in a recent epidemic, more than 700 000 hectares were affected. The larvae feed on a wide variety of conifers and deciduous trees; however, tree mortality is usually restricted to mature stands of the principal host trees, balsam fir and eastern hemlock.

The overwintering eggs hatch in late May or June and the young larvae feed on the new foliage. Later the larger larvae feed on the old needles. They are wasteful feeders and the whole needle is rarely eaten □. The damaged needles turn brown and the general browning of foliage in July indicates an abundance of feeding larvae. In late summer the mature larva □, about 30 mm long, ceases feeding and changes to a pupa □ in a protected niche on the tree or among debris on the ground. The moths □ emerge later in the fall and deposit groups of one to three eggs in a variety of locations including moss on the

Hemlock looper feeding injury

Eggs on lichen

Hemlock looper larva

forest floor, and lichens ☐ and bark scales on the stem and branches of trees. The eggs overwinter.

Hemlock loopers usually persist in an area for a few years in epidemic numbers. An abrupt reduction in numbers often follows as a result of various natural control factors including predators, parasites, disease, starvation and weather. The most important of these factors in the late 1960's epidemic appeared to be disease, particularly two fungi, *Entomophthora sphaerosperma* Fresenius and *E. egressa* McLeod & Tyrrell. Research is in progress to produce laboratory cultures of these fungi and to infect larval populations artificially early in their cycle of increasing numbers before feeding damage to the tree becomes serious. In spite of all these natural control factors, chemical insecticides will still be necessary in the future to control this pest. The application of large-scale control measures by forest managers is beyond the scope of this book. On the other hand, to reduce damage to

Hemlock looper pupa

Hemlock looper adult

specimen trees in infected areas, a contact or stomach insecticide recommended for caterpillars applied as soon as larvae are seen would be appropriate.

The full-grown larvae of the remaining loopers treated here vary in size from 15 to 35 mm. Some feed briefly in the fall and complete their feeding in the spring; others feed mainly in summer, while a third group feeds mainly in late summer and fall. Full-grown larvae are illustrated.

Loopers, fall and spring feeding

Fringed looper

Dashlined looper

The fall and spring feeding group includes the dashlined looper, *Protoboarmia porcelaria indicataria* (Walker); the diamondbacked looper, *Hypagyrtis piniata* (Packard); the fringed looper, *Campaea perlata* (Guenée); and the false loopers, *Syngrapha* species.

Diamondbacked looper

False looper

Loopers, summer feeding

False hemlock looper

The summer feeding group includes the false hemlock looper, *Nepytia canosaria* (Walker), occasionally found in large numbers and usually in conjunction with the hemlock looper; the small spruce loopers, *Eupithecia* species; the yellowlined conifer looper, *Cladara limitaria* (Walker); and the saddleback looper, *Ectropis crepuscularia* (Denis & Schiffermüller).

Saddleback looper

Yellowlined conifer looper

Small spruce looper

Loopers, late summer and fall feeding

Spruce looper

The late summer and fall group includes the commonly found spruce fir looper, *Semiothisa signaria dispuncta* (Walker) and the similar species *Semiothisa fissinotata* (Walker) on hemlock; the small spruce loopers, *Eupithecia* species (see illustration on page 69); the spruce looper, *Caripeta divisata* (Walker), occasionally abundant; the transversebanded looper, *Hydriomena divisaria* (Walker); and the whitelined looper, *Eufidonia notataria* (Walker).

No control measures have been required for these particular groups of loopers.

Spruce fir looper

Whitelined looper

Transversebanded looper

Hairy larvae

Rusty tussock moth

Three tussock moths occur commonly from Alberta to Newfoundland and in the northern United States. Larvae of the **rusty tussock moth**, *Orgyia antiqua nova* Fitch, feed on both coniferous and deciduous trees and large outbreaks have occurred on balsam fir and spruce in Newfoundland, usually in conjunction with the hemlock looper. The overwintered eggs hatch in the spring and feeding larvae can be found throughout June, July and August. When full grown the larvae □ are about 28 mm long. They change to pupae in yellow-grey cocoons in a variety of niches. The adults, winged males and wingless females, emerge mainly in August and September. Eggs are deposited in a single-layered mass on the cocoon.

Larvae of the **pine tussock moth**, *Dasychira plagiata* (Walker), have been reported on spruce, fir, hemlock and occasionnally jack pine. Although injury has been negligible in Canada, heavy feeding has occurred on jack pine in Wisconsin and Minnesota. After hibernation on the tree the young larvae begin feeding in the spring, and when full grown □, in June or July, are about 35 mm long.

Larvae of the **whitemarked tussock moth,** *Orgyia leucostigma* (J. E. Smith), feed on both deciduous and coniferous trees and young balsam fir has occasionally been damaged in the Maritimes. This species overwinters in the egg stage and the larvae may be found from spring to fall, but they are more abundant in July and August. When full grown □ they are about 35 mm long. The larvae change to pupae in a variety of niches in grey cocoons of silk and hair. The adults emerge in about 2 weeks and the wingless female, after mating with a winged male, lays her eggs in white masses on the cocoon. Winter is passed in the egg stage. In the Maritimes a naturally occurring virus was considered the key agent in ending recent epidemics of this species.

When necessary, a contact or stomach insecticide registered for use against caterpillars will provide adequate control against these tussock moths on valuable ornamental trees and in plantations.

Pine tussock moth

Whitemarked tussock moth

Naked larvae

Redmarked caterpillar

Variable climbing cutworm (green form)

The **redmarked caterpillar**, *Feralia jocosa* (Guenée), occurs in small numbers on various conifers from British Columbia to Newfoundland and in the northeastern United States. The larvae □ feed from about mid-May to late August and when full grown are about 30 mm long. The overwintering stage is the pupa.

The **variable climbing cutworm,** *Anomogyna elimata* (Guenée), with a brown and a green form □, is found from the Rocky Mountains in British Columbia to Newfoundland and in the eastern United States. It occurs in small numbers on fir, spruce, hemlock and other conifers. This species overwinters as a young larva and feeds the following spring until late June, when it attains a length of about 26 mm.

The **spruce harlequin,** *Palthis angulalis* (Hübner), is found in low numbers on spruce, fir and many other kinds of trees in North America. The larvae □, about 20 mm long when full grown, may be found from May to the end of September.

The **fir harlequin,** *Elaphria ver-*

Spruce harlequin

Fir harlequin

sicolor (Grote), feeds on balsam fir, spruce, hemlock and other tree species from Manitoba to Newfoundland and south to Georgia. No injury to trees has been reported. The larvae □, about 20 mm when full grown, may be found from June to the end of September.

No control measures have been required for this group of moth larvae.

Sawfly larvae

*Head colour is yellow on mature larvae and black on earlier larvae.

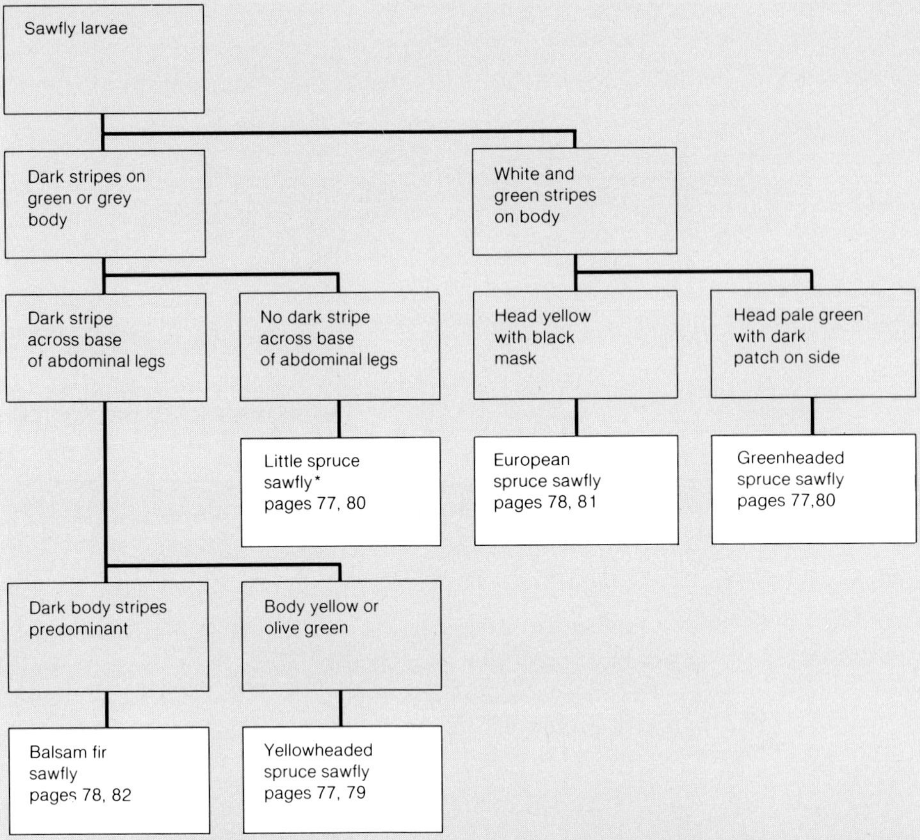

Yellowheaded spruce sawfly

Greenheaded spruce sawfly

Little spruce sawfly

Balsam fir sawfly

European spruce sawfly

Frontal view of larva

Yellowheaded spruce sawfly and others

The yellowheaded spruce sawfly, *Pikonema alaskensis* (Rohwer), is a ubiquitous pest of spruce throughout most of Canada and the northern United States. Because young, open-grown trees are preferred, plantations, prairie shelterbelts, hedges, roadside plantings and ornamental trees often suffer severe feeding injury, and on occasion open-growing young trees in forest stands are also damaged☐.

This insect overwinters as a larva in the ground in a dark-brown papery cocoon encrusted with soil☐. In the spring it changes to a pupa and the adult☐ emerges mainly in May or June but occasionally later in the summer. The spring-emerging females lay their eggs one per needle in a shallow slit at the base of the young needles☐ about the time the bud cap is shed. The eggs hatch in 7 to 14 days and the young larvae feed in groups on the new needles until only short, brown stubs are left. When the new growth is eaten, the larvae move back on the twig to feed on the older needles until they are full grown and about 20 mm long (☐ page 77). Feeding is usually completed in July, at which time the larvae drop to the ground and overwinter in the soil inside a cocoon.

In spite of the large numbers of parasites that attack this sawfly, chemical control measures have frequently been necessary to protect planted trees. When required, an insecticide suitable for sawflies should be applied about 10 days after the bud caps have been shed. This should ensure a minimum amount of damage and the greatest larval mortality. When only a few trees are involved,

Heavily damaged young tree

Soil-encrusted cocoon

the larvae can be handpicked and destroyed.

The closely related **greenheaded spruce sawfly**, *Pikonema dimmockii* (Cresson), has a distribution similar to that of the foregoing species but occurs most commonly from Saskatchewan eastward. Noticeable feeding injury caused by this sawfly is rare. The larva (□ page 77) is about 20 mm long when full grown. It has a life history similar to that of the more destructive yellowheaded spruce sawfly, and is often found along with it.

The **little spruce sawfly**, *Pristiphora lena* Kincaid, is a little-known species that has been reported from Ontario and Newfoundland. Serious feeding injury has not occurred. The larvae feed on spruce in June and July in Ontario and are black headed until they are almost full grown, when the head colour changes to reddish brown. The full-grown larva (□ page 77) is about 10 mm long and, like other sawflies on spruce, it also overwinters in a cocoon in the soil.

Adult

Eggs

European spruce sawfly

Larva

Cocoon

Adult

The European spruce sawfly, *Gilpinia hercyniae* (Hartig), was first found in North America in Quebec, near Hull, in 1922. By 1930, severe damage was occurring in spruce stands on the Gaspé Peninsula and in the succeeding decade further damage occurred in New Brunswick and the northeastern United States. Extremely high populations followed, and more than 10 million cubic metres of wood were lost through tree mortality. The pest now occurs in relatively low numbers on all species of spruce from Manitoba to the Atlantic Ocean and in the adjacent parts of the United States.

There are one or two generations of this sawfly each year and the two generations have the following seasonal occurrence in Ontario. The larvae overwinter in cocoons in the litter layer and change to pupae in the spring. The adults emerge shortly afterward. Males are very rare and, as a result, reproduction usually takes place without fertilization. The females lay their eggs in slits cut in the needles

Balsam fir sawfly

and, upon hatching, the larvae feed on the older needles in June and July. When the larvae are full grown☐ and about 20 mm long, they drop to the ground and spin cocoons☐. Within a month the adults☐ emerge and lay their eggs. The second-generation larvae feed from mid-August through September.

The European spruce sawfly story is an excellent example of biological control. For the past 30 years there has been no need for chemical control of this pest, since the combined action of an accidentally introduced virus disease and of introduced parasites appears to be sufficient to maintain populations at low levels. The introduction of the masked shrew, a cocoon-hunting rodent, to Newfoundland has added another control factor there.

Damaged needles

The balsam fir sawfly, *Neodiprion abietis* complex, occurs from Alberta to Newfoundland and in United States from Missouri and the Lake States to New England. In Canada severe periodic epidemics, each usually of short duration, are common on balsam fir and spruce from Saskatchewan eastwards. Although some tree mortality has occurred, it is usually reported in conjunction with damage by the spruce budworm, the blackheaded budworm or the balsam woolly aphid.

Larva and cocoon

The balsam fir sawfly complex is apparently made up of four "strains" in the geographical area dealt with in this handbook, two strains on balsam fir and one each on white and black spruce. Because there is considerable difference in seasonal occurrence of the larval stages of the four strains, only a generalized life history is given below.

The overwintering eggs hatch in the spring and larvae of the various strains feed from mid-May to early August. At first they eat only part of a needle □ but, as they grow larger, the entire needle is consumed except for a stub. Young larvae feed in groups, but the older ones feed separately. The full-grown larvae (□ page 78), about 17 mm long, spin light-brown cocoons □ on the needles or in the ground litter, where they change to pupae. The pupae become adults in the fall and, after mating, the females lay their eggs in slits cut in needles.

Epidemics of the balsam fir sawfly have usually been of short duration and were probably terminated by a naturally occurring virus disease as well as by parasites. No large-scale control measures have been attempted. Larvae on small shade trees can be handpicked; on larger trees a contact or stomach type of insecticide applied as soon as the larvae are seen will control this pest.

Spruce webspinning sawfly

Nest of excreta

There are a number of webspinning sawflies that feed on spruce, but most of them are rare and are solitary feeders. The most common species is the spruce webspinning sawfly, *Cephalcia fascipennis* (Cresson), which occasionally causes damage to ornamental spruce trees or hedges across Canada and in the northeastern United States. The following comments relate to the above species.

The larvae overwinter in cells in the ground and change to pupae in spring. The adults □ soon emerge and the female lays green cylindrical eggs singly or in end-to-end rows of two to four on the needle. On hatching, the larvae construct shelters of silk and excreta □ at the crotch of a twig and branch. The larvae forage from these shelters by cutting the older needles and eating them from the base outwards. Tips of needles are usually not eaten and may be incorporated into the shelters. The larvae of this family typically have elongate antennae, lack abdominal legs and bear a pair of jointed appendages at the posterior end □ . Colour is variable, but the head is usually dark and the body brownish with a reddish line on the back. When full grown and about 25 mm long, the larvae drop to the ground to overwinter.

If control is required on ornamentals, nests containing larvae may be removed by hand and destroyed; a contact insecticide applied with sufficient force to penetrate the nest should also prove effective.

Larva

Adult

Needle miners

Spruce needleminer

European spruce needleminer

More than 10 species of needle miners have been found on the treated group of trees. The types of injury caused by the most common species are shown here and information on them is given in the succeeding five pages.

Orange spruce needleminer

Hemlock needleminer

Spruce budworm

Spruce needleminer

Spruce needleminer

The spruce needleminer, *Endothenia albolineana* (Kearfott), is widely distributed in Canada and the United States. It is primarily a pest of planted spruce and has not been a serious problem in the forest.

The nearly fully-grown larvae overwinter in nests of dead, mined needles and excreta tied together loosely with silk (☐ page 86); some may overwinter in mined needles. In the spring the larvae resume feeding by tunnelling inside the needles. The full-grown larvae ☐, about 8 mm long, change to green pupae in grey cocoons in the nest of dead needles from May to July. The pupae become adults in about 2 weeks and the females lay their eggs on the needles in groups of 3 to 10. On hatching, the new generation of larvae mine the older needles from summer into October, when they go into hibernation among the dead, mined needles.

Although the feeding by this needleminer will not kill a tree, it often makes an ornamental tree look unsightly. If the nests of dead needles cannot be pruned and destroyed in early spring, a systemic type of insecticide should be applied either in early May or mid-August, or during both months.

European spruce needleminer

European spruce needleminer

The European spruce needleminer, *Epinotia nanana* (Treitschke), is known to occur in British Columbia, Ontario and Quebec, and in the United States from Maine to Ohio and Michigan. To date it has been a pest mainly of planted trees. Although the larvae feed on most species of spruce, Norway spruce appears to be preferred.

The nearly full-grown larvae overwinter in mined needles and resume feeding in early spring in adjacent needles. Each larva mines from 6 to 10 needles in the spring (□ page 86) and secures them to the twig with silk. The full-grown larva □, about 9 mm long, changes to a pupa in a silken cocoon in the soil litter or occasionally on the tree, in May or June. The pupae become moths in 4 weeks and the females lay eggs, usually singly, occasionally in clusters, on needles produced in the previous year. The eggs soon hatch and the larvae feed in needles during summer and fall and hibernate in the needle mine.

When shade trees are made unsightly by this needleminer, a systemic insecticide applied in late April and/or during August should provide control.

Orange spruce needleminer

Orange spruce needleminer

The orange spruce needleminer, *Coleotechnites piceaella* (Kearfott), is found on spruce and balsam fir from Alberta to Nova Scotia and in the northeastern United States. No serious feeding injury is recorded.

The larvae overwinter in silken shelters at the base of mined needles or in dead foliage (☐ page 87). Feeding is resumed in the spring inside the previous year's needles or in swollen buds. When full grown in late June or early July, and about 8 mm long, the larvae ☐ change to pupae in silken cells on the foliage or on the ground. The pupae become adults in about 12 days and the females lay their eggs in the foliage. The eggs hatch and the young larvae feed from August to late September before hibernating near the feeding sites. The fall feeding is in mined needles although larvae are also found in dead foliage. It is believed that in the dead foliage they feed on fungi.

The most effective chemical control agent against this insect would be one with systemic activity and it should be applied when the buds begin to swell.

Hemlock needle miners

Green hemlock needleminer

There are two closely related species of needle miners on hemlock. Except for colour they are similar in most respects. The green form is the most common one in Ontario.

The **green hemlock needleminer,** *Coleotechnites apicitripunctella* (Clemens) occurs on hemlock in eastern Canada and local outbreaks have been reported from the northeastern United States. The larva overwinters in the needle mine and resumes feeding in the spring. Adjacent needles on a twig are mined and tied loosely together with silk (☐ page 87). As the larva approaches maturity it hollows out needles from the underside. The full-grown larva ☐, about 6mm long, is green with a pale-brown head. In late May or June it changes to a pupa in a silk tube that connects the mined needles.

The **brown hemlock needleminer,** *Coleotechnites macleodi* (Freeman), has a brown body and a black head.

Control measures have not been required for the hemlock needle miners. If they should become necessary on ornamentals, a systemic insecticide applied during the larval stage would be appropriate.

Spruce budworm

Immature spruce budworm

In addition to the preceding needle miners, which spend most of their feeding days inside a needle, the larvae of a number of insects tunnel in the needles for a short while when they are quite small and later feed externally. The most common of these insects is the spruce budworm, *Choristoneura fumiferana* (Clemens). It overwinters as a young larva in various hibernation sites on balsam fir or spruce. In early spring the hungry larvae spin a silken web, usually between two needles of the previous year's growth, and start eating their way into one needle (☐ page 87). At this stage they are light brown with black heads ☐. When the buds start to elongate, the larvae cease their mining habit and move to feed on the new needles. The complete life cycle of the budworm is explained on pages 54 to 59.

Balsam gall midge

*New information indicates that larvae of two species may be present. The actual gall-former is *Paradiplosis tumifex* Gagné.

The balsam gall midge, *Dasineura balsamicola* (Lintner)*, is probably present throughout most of the range of balsam fir in North America. The early shedding of galled needles has been of particular concern to Christmas tree growers in the Maritime Provinces and in the northeastern United States, where balsam fir comprises a large proportion of their crop.

The mature larvae overwinter in the soil and change to pupae in early spring. The adult midges emerge as the new shoots are beginning to develop and lay their eggs between the young needles. The feeding of the young larva causes the needle tissue to envelop the larva □ and, when midges are abundant, three or more galls on one needle are not uncommon. The orange larvae feed inside the galled needles until late September or October. When full grown and about 3 mm long, they drop to the ground, where they spend the winter.

The balsam gall midge is probably a serious pest only in young balsam fir forests or plantations, where the trees are destined for the Christmas tree trade. Where large-scale control measures may be required, guidance, especially on the timing of application of a suitable insecticide, should be sought from the appropriate Forest Research Centre listed on page 12. On specimen trees needle losses in 1 year are lost sight of in subsequent years as new growth is produced; hence, the application of an insecticide is seldom warranted.

Galled needles

Spruce spider mite

Adult mite

Normal and damaged foliage

The spruce spider mite, *Oligonychus ununguis* (Jacobi), occurs throughout much of Canada and the United States. Although it does not appear to be important in the forest, it has been particularly bothersome on planted spruce in the Prairie Provinces and is considered one of the most important pests of spruce plantations in Ontario. In addition, ornamental spruce and most other conifers elsewhere throughout its range are often injured. The pest, which is barely visible without the aid of some magnifying device, is rarely noticed until discoloured foliage □ is found on a tree. To determine if mites are the cause of damage, suspected foliage should be sharply jarred over a piece of white paper and the paper watched for the movement of tiny objects. If, when the objects are pressed and brushed against the paper, a red smear results, it is highly probable that spruce spider mites are present.

Spruce spider mites suck sap from the needles and shoots, causing the foliage to take on a mottled appearance. They also spin fine silk webbing

Eggs on cedar

as they move about the needles; dust particles in this webbing add to the unhealthy appearance of the trees. The mites tend to become particularly numerous and destructive during hot dry seasons. They are readily dispersed by wind and may also be spread through the shipment of infested nursery stock.

This pest overwinters in the egg stage, mostly on the twigs. In the Prairie Provinces the eggs hatch in May. The young, which to the layman resemble the adult mite □, develop into first-generation adults by early June. By mid-June the female, only about 0.5 mm long, lays 40 to 50 eggs on the foliage □. These eggs give rise to a second generation of adult mites in late June. Three to six generations follow throughout summer and early fall. The overwintering eggs are laid from September to the onset of severe frost.

Chemical control measures are frequently necessary to prevent damage to ornamental trees, and pesticides effective against mites should be used. To prevent damage by these mites, trees exhibiting off-colour foliage should be checked throughout the summer.

Pine needle scale

The pine needle scale, *Chionaspis pinifoliae* (Fitch), is a common pest of pine, spruce and other conifers across Canada and throughout the United States. In the Prairie Provinces it has been a particularly persistent pest on planted spruce in both rural and urban settings. In heavy infestations the needles appear to be flecked with white □. This insect overwinters in the egg stage under the white covering of the scale □. In Saskatchewan hatching dates vary from late May to late June. The newly hatched "crawlers" disperse and settle on old or new needles and scale development begins. For further information on this insect see the earlier handbook in this series, *Insects of Eastern Pines.*

The **hemlock scale**, *Abgrallaspis ithacae* (Ferris), is an uncommon scale on the underside of the needles of spruce, fir and hemlock. The scale is oblong in outline and dark coloured.

Chemical control measures are most effective during the crawler stage of most scales and a contact insecticide should be used.

Heavy scale on spruce

Scales on needles

Undamaged white spruce needles on new shoots

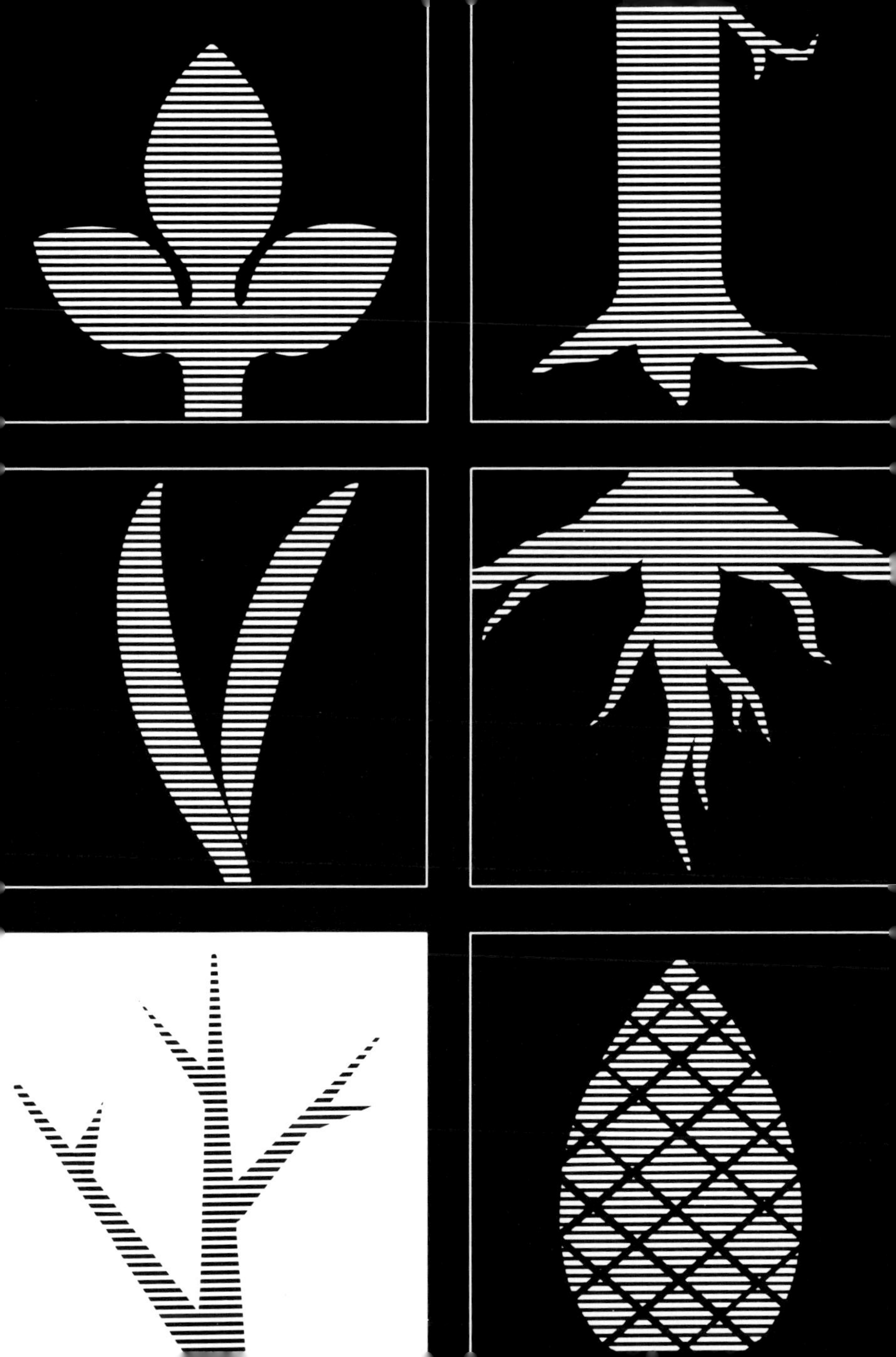

Twig or branch

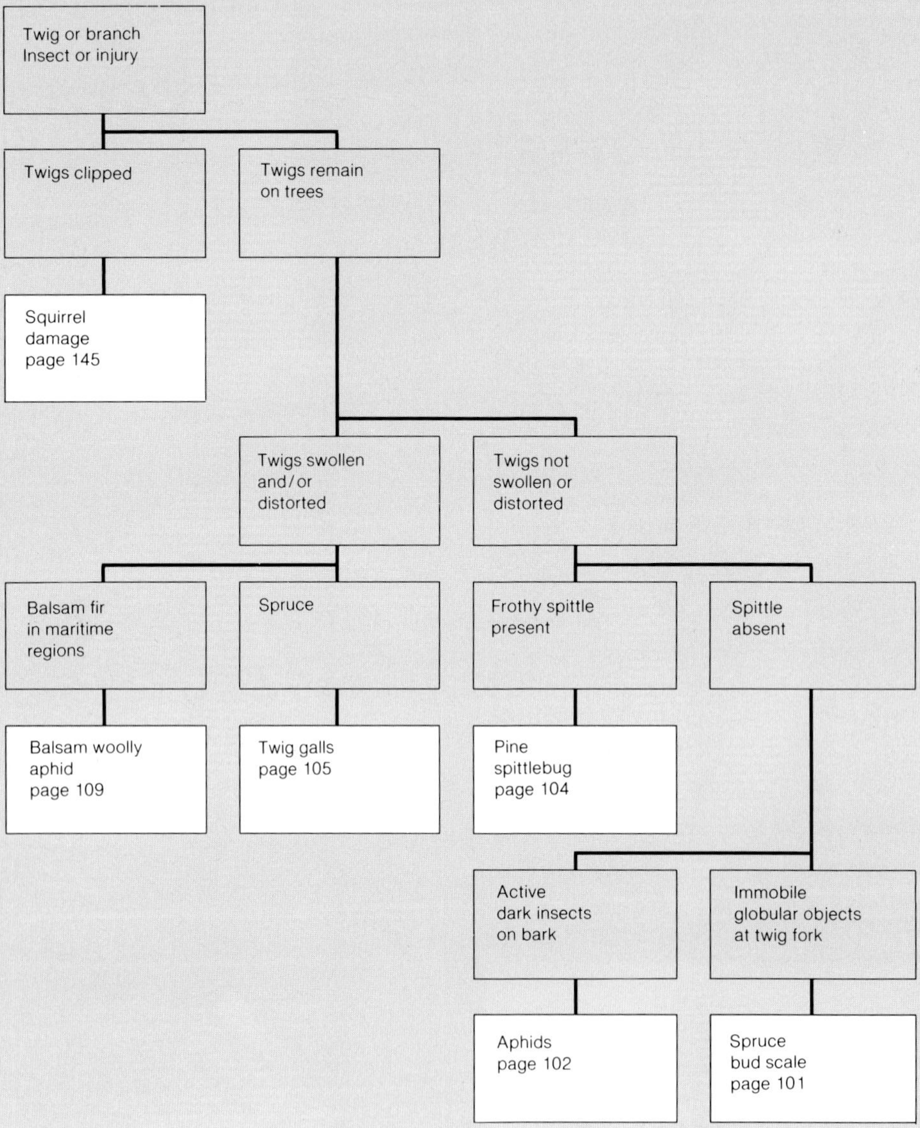

Spruce bud scale

The mature spruce bud scale, *Physokermes piceae* (Schrank), which is frequently mistaken for a bud, is known from the Northwest Territories to the Maritime Provinces and in the north central and northeastern United States. Although it feeds on the native spruces and occasionally on balsam fir, it is most often abundant on Norway spruce. Recently, large, severe infestations have occurred in spruce plantations in New Brunswick. Heavily infested trees are weakened by the feeding, which causes needle loss, and are made unsightly by a black mould that develops in the liquid excretion of the scale insects. This excretion is attractive to both bees and ants, whose presence often draws attention to the scale.

The scale develops through one generation each year and overwinters as immature individuals clustered around terminal buds, their feeding tubes inserted in the sap stream. They resume feeding in the spring and mature in early summer. At this time the female scales are about 3 mm in diameter □. The eggs are formed under the scale. As they hatch, the young crawlers move out to the tips of twigs, where they settle, insert their fine feeding tubes and feed until late fall.

When control of the scale is necessary, a systemic insecticide applied in early August, or a contact insecticide applied when the crawlers are active, from July to early August, should prove effective.

Mature scales

Aphids

Aphid colony

A number of species of aphids of the genus *Cinara* feed on spruce and fir in North America, but noticeable injury by these aphids is unknown. They pierce the bark with their long feeding tubes and feed on sap from shoots, twigs, branches, stem and roots. Most species feed in groups ☐ and are usually attended by ants, which feed on the droplets of excreted liquid. These aphids vary in colour from grey to brown or black and are less than 5 mm long ☐. All species overwinter in the egg stage. The eggs are blackish and are laid singly or in rows on the needles ☐. Six generations in 1 year are not unusual in Canada and succeeding generations often move to new sites on the tree, including the roots, as the season progresses. The life cycle is complex; e.g., adults of the intermediate summer generations consist of females only, some winged and others wingless, which produce tiny nymphs rather than eggs. Males occur only in the late fall generation, which produces the overwintering eggs.

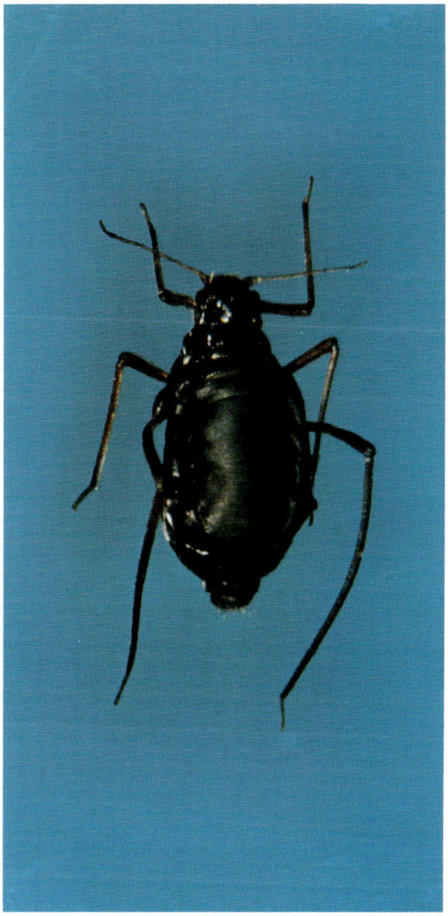
Wingless aphid

Eggs

No large-scale control measures have been required for these aphids. When control on ornamentals is necessary, a contact insecticide would be appropriate.

Pine spittlebug

Spittle mass on balsam fir

Bug in spittle mass

The pine spittlebug, *Aphrophora cribrata* (Walker) occurs throughout central and eastern Canada and the eastern United States. In Ontario it is of concern mainly in pine plantations and, although it is common on fir and spruce, no instances of serious injury to these trees are known. Initially, frothy masses are seen near the tips of the branches. As the season progresses, they are observed closer to the main stem. Under each frothy mass of spittle □ will be found one or more immature spittlebugs □ feeding from May to July. For control, apply a contact insecticide in June with sufficient force to penetrate the spittle mass. Additional information on this insect may be found in *Insects of Eastern Pines*, an earlier handbook in this series.

Twig galls

Ragged spruce gall adelgid injury

Injury to twigs of spruce by the **ragged spruce gall adelgid,** *Pineus similis*(Gillette), persists from year to year ☐ and soon results in unsightly trees. For information on this and related species see "Gall insects", on pages 42 to 48.

Old injury by the **spruce gall midge,** *Mayetiola piceae* Felt, is also unsightly ☐, but large numbers of this insect are uncommon. More detailed information on this midge is given on page 48.

Spruce gall midge injury

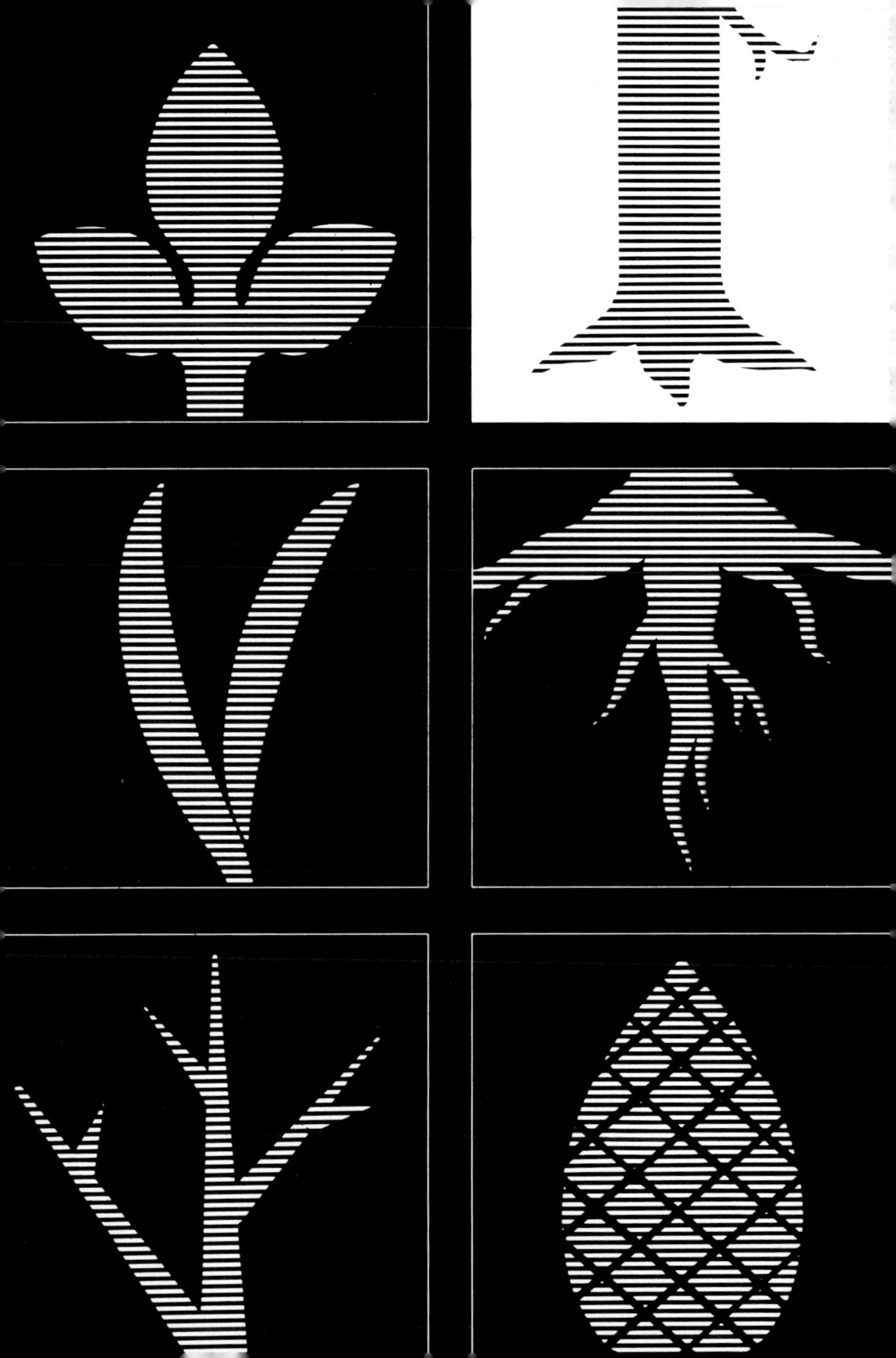

Stem or log

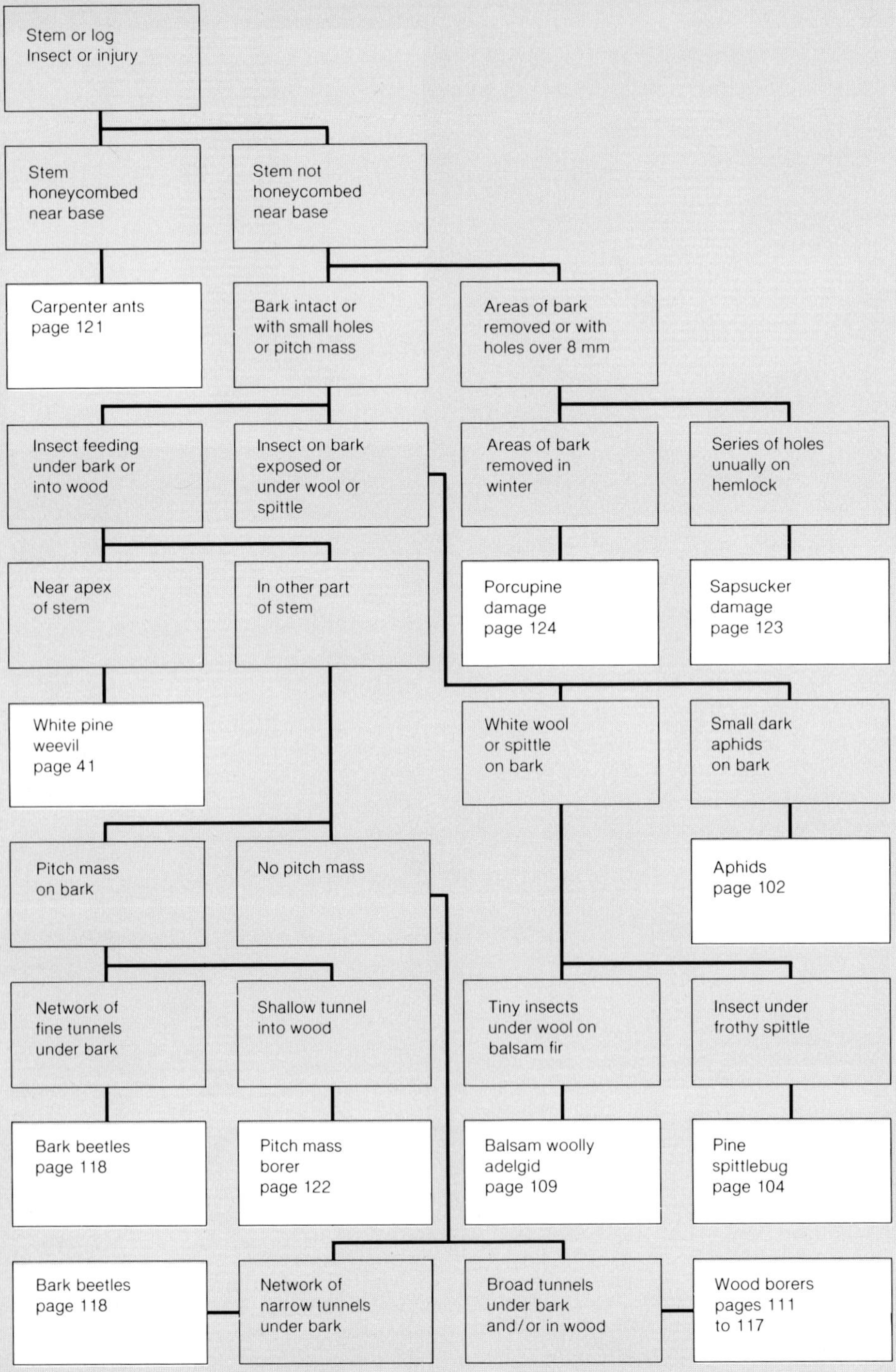
Stem or log Insect or injury
Stem honeycombed near base
Stem not honeycombed near base
Carpenter ants page 121
Bark intact or with small holes or pitch mass
Areas of bark removed or with holes over 8 mm
Insect feeding under bark or into wood
Insect on bark exposed or under wool or spittle
Areas of bark removed in winter
Series of holes unually on hemlock
Near apex of stem
In other part of stem
Porcupine damage page 124
Sapsucker damage page 123
White pine weevil page 41
White wool or spittle on bark
Small dark aphids on bark
Pitch mass on bark
No pitch mass
Aphids page 102
Network of fine tunnels under bark
Shallow tunnel into wood
Tiny insects under wool on balsam fir
Insect under frothy spittle
Bark beetles page 118
Pitch mass borer page 122
Balsam woolly adelgid page 109
Pine spittlebug page 104
Bark beetles page 118
Network of narrow tunnels under bark
Broad tunnels under bark and/or in wood
Wood borers pages 111 to 117

Balsam woolly adelgid

* This account and illustrations were kindly supplied by R. S. Forbes of the Maritimes Forest Research Centre.

The balsam woolly adelgid* □, *Adelges piceae* (Ratzeburg), is of European origin and was first found in North America in Maine in 1908. It is one of the more important pests of balsam fir and the other true firs on this continent. It occurs in Newfoundland, the Maritime Provinces, eastern Quebec, southwestern British Columbia, and in the northeastern and northwestern United States.

The insect feeds on cell sap within the bark cortex from all parts of the host tree from the root collar to the crown, causing two types of damage. One is on the stem where more than 3 adelgids per cm^2 of bark surface □ can kill trees in 3 years. The other is in the crown where high populations at the nodes and bases of young shoots may cause pronounced swelling of shoots □, branch mortality, impairment of height growth, and eventual death of trees. Larger trees in a stand are generally infested first, but smaller trees and reproduction often become infested as well, and may die □. In the Maritime Provinces, where the adelgid has been present longer than anywhere else in Canada, large quantities of balsam fir have been killed and much growth has been lost. Indeed, through growth stagnation, the adelgid has made this tree almost a weed in some areas.

The black, tiny, young adelgids overwinter on the stem or at the base of twigs with their mouthparts inserted into the bark. In the spring, they begin to feed and, after three moults, become adults, usually about the time of bud burst. The wingless adults are about 1 mm long and lay pale yellow eggs which later become light purplish

Heavy stem attack

Balsam woolly adelgid

Gouty twig

Trees killed by stem attack

brown □. The eggs give rise to nymphs which wander considerably before settling on suitable feeding sites. Once settled, the nymphs turn black and remain stationary for up to 4 weeks before beginning to feed. There are two to four generations, depending on summer temperatures. There are no males.

Apart from temperatures below −32°C, which regulate distribution, population level and the amount of damage caused, there are no known controls entirely satisfactory for forested areas. Introduced predators offer reasonable control of stem populations but are ineffective against twig populations. Chemical insecticides are not practical for protection of forests, but systemic insecticides may prove useful in protecting high-value trees. Spread of infestations may be checked and losses minimized by salvaging infested stands in winter and clear-cutting and burning slash. Also recommended, at least in eastern Canada, are cutting fir on short rotations and reducing its content in stands.

Adults, eggs, and white wool

Wood borers

Roundheaded borer tunnel

Flatheaded borer tunnel

Horntail tunnel

Wood borers are usually abundant in damaged or decadent forests, or in logs that are left in the forest during the summer. In the undisturbed forest, the role of borers is beneficial in that they hasten the breakdown of dead trees by opening up the wood to wood-rotting fungi. However, when forests are extensively injured by wind, fire, insects or other agents, attempts to salvage merchantable wood are often hampered by borers, which riddle the dead or dying trees with their larval tunnels. These produce the common lumber defect known as "wormholes". Three families of insects cut tunnels into the wood. Typical tunnels of a roundheaded borer □, a flatheaded borer □ and a horntail □ are shown, and further information is given on the following pages.

Roundheaded borers

Whitespotted sawyer

There are at least 20 species of roundheaded borers, Cerambycidae, that feed on the wood of spruce, fir and hemlock. Only exceptionally are borers found tunnelling in living trees, although feeding by the adults on the tender bark of twigs when populations are high may cause damage to young living trees. The **whitespotted sawyer,** *Monochamus scutellatus* (Say), is one of the most common and widely distributed borer species in North America. The robust adults □, seemingly clumsy fliers, are present during the summer, when they may be found on newly fallen or recently cut trees chewing tiny slits in the bark in which they lay eggs □. The eggs hatch in about 2 weeks and the tiny larvae tunnel to the wood and score its surface with their feeding. With the advent of cooler weather, they enter the wood through oval entrance holes and tunnel deep into it. Feeding continues the following summer, when larvae □ occasionally return to the surface of the wood, and the tunnels are extended generally in a U-shaped

Egg greatly enlarged

Mature larva

configuration. It is during this time that small piles of fibrous material extruded by the larvae accumulate under logs □. Early in the spring of the second year following egg laying, the larvae, when they are about 30 mm long, change to pupae in the tunnel enlargement just below the wood surface (see page 111). In early summer the resulting adults chew their way out, leaving round exit holes (also shown in the tunnel illustration) as evidence of emergence, completing the usual 2-year life cycle.

Because of the serious downgrading of lumber produced from logs attacked by borers □, it is frequently necessary to apply insecticides to logs in the bush to prevent feeding. For information on the control of borers in logs it would be advisable to consult the appropriate research centre listed on page 12. Kiln drying will kill tunnelling larvae in lumber. Adults emerging from lumber or logs in completed structures will not reinfest the material. Control of most roundheaded borers in the stems of damaged or dying

Piles of fibrous material

Lumber damaged by *Tetropium* species

shade trees is seldom warranted, because the trees are probably already doomed.

Flatheaded borers

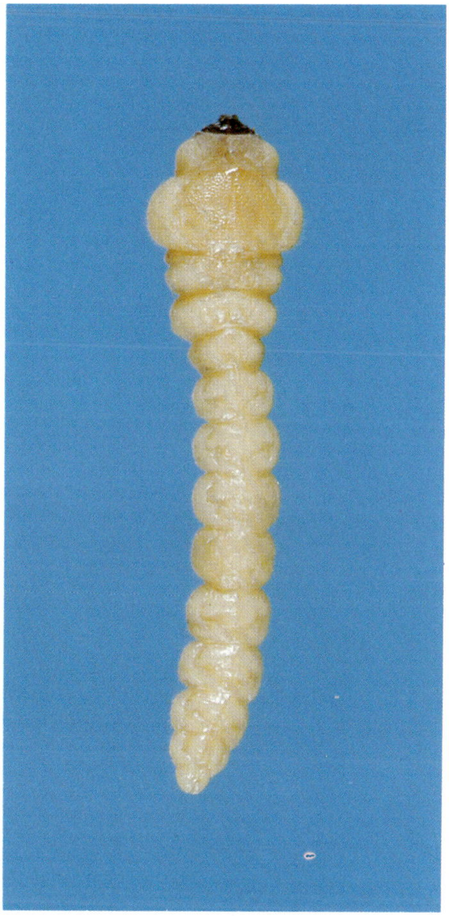

Larva

Adult

Of the 10 species of flatheaded borer, Buprestidae, which feed on spruce, fir and hemlock, most are found on hemlock. As with the roundheaded borers, most feeding occurs in dying or dead trees or near injuries on living trees. Damage becomes abundant only where a continuing supply of breeding material is available. The life history of these borers is similar to that of the roundheaded borers, 2 years being required to complete the cycle. However, under adverse conditions some exceedingly long life cycles have been reported. The full-grown larvae may be up to 25 mm long. Characteristically they are flattened, the anterior part of the body being much broader than the remainder □. The bronzed adults □ are usually seen only where suitable material occurs in sunny locations.

Control of flatheaded borers is not usually required, but if it does prove necessary, they should be treated in the same way as the roundheaded borers.

Horntails

Mature larva

Members of three genera of horntails, Siricidae, attack dying or recently killed balsam fir and spruce. The common name is derived from the stout spine-like structure at the end of the adult's body □, which is used to pierce the bark and allow the insertion of the eggs in the wood. As in the case of the preceding beetle borers, rapid population increases occur only where brood material is quite abundant. The life cycle usually requires 2 years, but the adults are in flight later, usually from late July to early September. The full-grown larva □ may be up to 25 mm long. Wood rotting fungi found in horntail tunnels cause very rapid deterioration of wood and, consequently, reduction of the time during which wood can be salvaged. Large-scale control measures have not been necessary to date. Under normal circumstances the larvae are preyed upon by a spectacular-looking, wasp-like parasite that can detect the presence of a horntail larva tunnelling in the wood. This parasite is capable of inserting its ovipositor, a sheathed, hair-like structure that may be up to

Urocerus albicornis adult

Parasite of the genus *Megarhyssa*

10 cm long, through bark and wood into the tunnel. An egg is deposited and the resulting larva feeds on the horntail larva. The parasite □ is in the process of inserting its ovipositor into a horntail tunnel in a maple tree.

Bark beetles

Balsam fir bark beetle

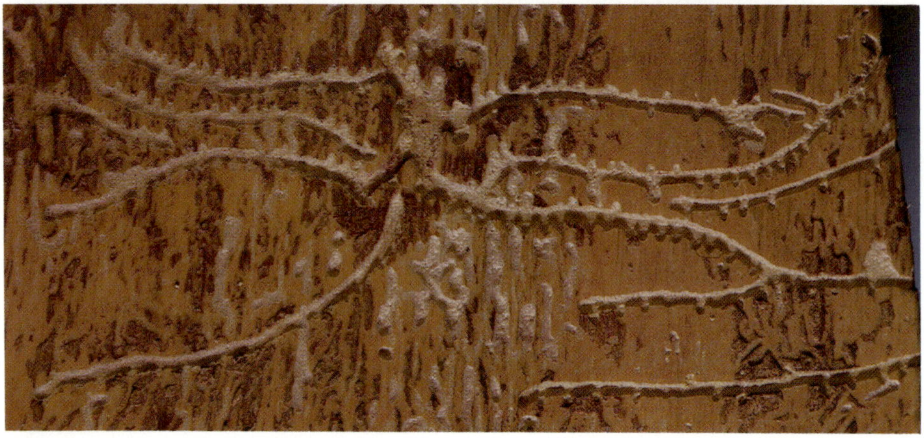
Gallery

Bark beetles, Scolytidae, feed and breed between the bark and the wood, usually engraving both with their galleries. More than 20 different kinds of these tiny beetles feed on weakened, dying or dead spruce, fir and hemlock. Most of them restrict their breeding area to one part of the tree, i.e. twig, branch, stem or root collar; some breed in only one kind of tree whereas others breed in a wide range of conifers. Since they frequently breed in trees initially damaged by other means, they are generally considered to be secondary pests and in undisturbed eastern forests may be thought to be beneficial in that their feeding hastens the return of wood to the humus.

However, where extremely high populations occur as a result of large supplies of breeding material, healthy trees nearby may succumb to sustained severe attack. Because of the number of species involved and their similarity in all stages, identification is best left to specialists. Two of the most important species will be dealt with briefly to characterize typical life histories. Both species have occurred in outbreak numbers after spruce budworm epidemics.

Overwintered adults of the **balsam fir bark beetle** □, *Pityokteines sparsus* LeConte, about 2.5 mm long, select suitable trees in early summer and form nuptial chambers under the bark, each one containing a number of females and one male. When these have established a series of tunnels containing eggs for one brood □, they move to a second tree to create a second brood later in the summer. Progeny of brood 1 reach the adult stage in August and some of them move to a new tree. Progeny of brood 2 pass the winter as larvae and change to the pupal and adult stages early the following summer. As a result, all stages can be found through much of the summer. This is the most common species found on balsam fir.

The **spruce beetle**, *Dendroctonus rufipennis* Kirby (formerly *D. obesus* Mannerheim), about 6 mm long, is one of the larger bark bettles □ found in spruce. Initial attack is in the lower trunk and is indicated by the red boring dust in the bark crevices and by pitch tubes □ , especially when weakened or recently dead trees are attacked. Overwintering adults construct egg tunnels for brood 1 in June and a second set of tunnels in late July for brood 2. Some members of brood 1 emerge as adults in late July and construct additional tunnels, while others overwinter as mature larvae □ and emerge as adults in July along with another segment of the population that has overwintered as early

Spruce beetle

larvae. A smaller bark beetle often found in spruce, because of its divided eyes, is named the **foureyed spruce bark beetle**, *Polygraphus rufipennis* Kirby.

The removal of brood material from the forest will prevent the development of destructive populations, but if control of bark beetles in natural stands is necessary, it is best left to specialists. On the other hand, control of bark beetles in individual shade

Spruce beetle larva

Pitch tube

trees is seldom warranted, since trees susceptible to attack are probably already doomed as a result of other factors.

Carpenter ants

Carpenter ants, the largest of our common ants, are often found in the lower trunk of living trees or in logs with unsound heartwood, and on occasion in log structures or in the framing members of buildings. Wood containing an ant nest is honeycombed with many chambers □. Although tunnels may initially be in unsound wood, they are often extended into sound wood. The ants do not eat wood but simply remove it to construct chambers. They feed mainly on dead and living insects, sweet compounds and refuse.

Colonies start with a fertilized female in a suitable cavity, where she rears an initial brood of small-sized workers. Subsequent broods, which are fed by the workers, are larger. Long-established colonies eventually contain a reproductive female, winged males and virgin females as well as workers □, sometimes in thousands. After a nuptial flight in early summer, the mated female seeks a suitable nest site, sheds her wings, and establishes a new colony.

Control in the forest is not feasible. In logs and structural timbers, avoid unsound wood and treat wood in damp situations with preservatives. Use a contact insecticide dust on ant runways to destroy established colonies.

Camponotus herculeanus (worker)

Honeycombed wood

Pitch mass borer

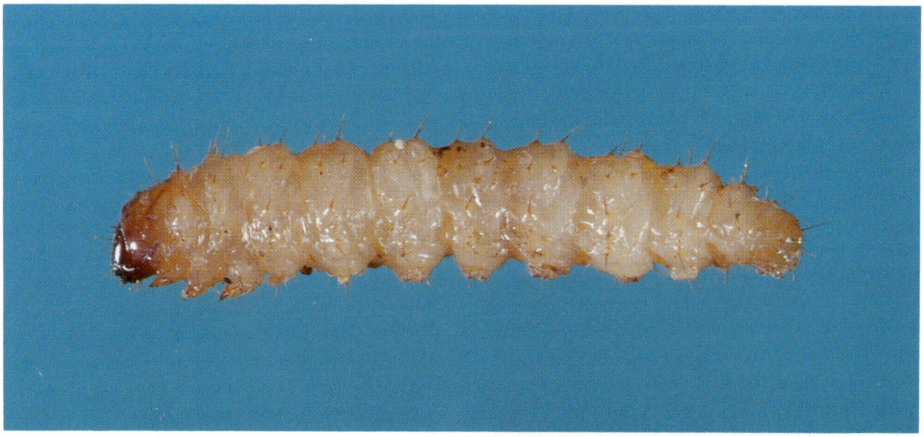

Larva

The pitch mass borer, *Synanthedon pini* (Kellicott), occurs on spruce and pine in eastern North America. It does not kill trees, but the pitch-filled larval tunnels in the wood cause defects in the lumber.

The adults are unusual in that the wings are mostly clear and therefore not typical of most moths. The eggs are laid on the bark in early summer, usually near a wound or in scars or crevices. The larvae □ feed in tunnels on the inner bark and sapwood, causing a copious flow of pitch, which hardens on the bark □ . They feed for 2 or 3 years and attain a length of about 25 mm before changing to pupae in the pitch mass. Adults emerge in early summer.

No control measures have been necessary in the forest. On ornamental trees the larvae may be removed with a knife and the wound allowed to heal.

Pitch mass

Sapsucker damage

Feeding injury

The yellow-bellied sapsucker, *Sphyrapicus varius varius* Linnaeus, a migratory bird and a member of the woodpecker family, feeds extensively on insects but also on the sap and bark tissues □ of living trees. The preferred tree species, birch and hemlock, are often severely injured and, when feeding results in girdling of the tree trunk, the part above is killed. Prevention of injury is almost unattainable.

Other members of the woodpecker family often chisel out larger holes from dead or dying trees in their search for various woodboring larvae discussed on pages 111 to 117.

Porcupine damage

Feeding injury

The porcupine, *Erethizon dorsatum* Linnaeus, feeds on a wide range of plants and the bark of many kinds of trees. Damage to trees □ occurs in areas near dens where a group is sheltered for the winter either in hollow logs or in caves among rocky outcrops. Although trees are often killed by the feeding on the bark, only a small number of trees are usually involved in any one locality.

Undamaged tree stems of: White spruce

Colorado spruce

Black spruce

Balsam fir

Norway spruce

Eastern hemlock

Root collar or root

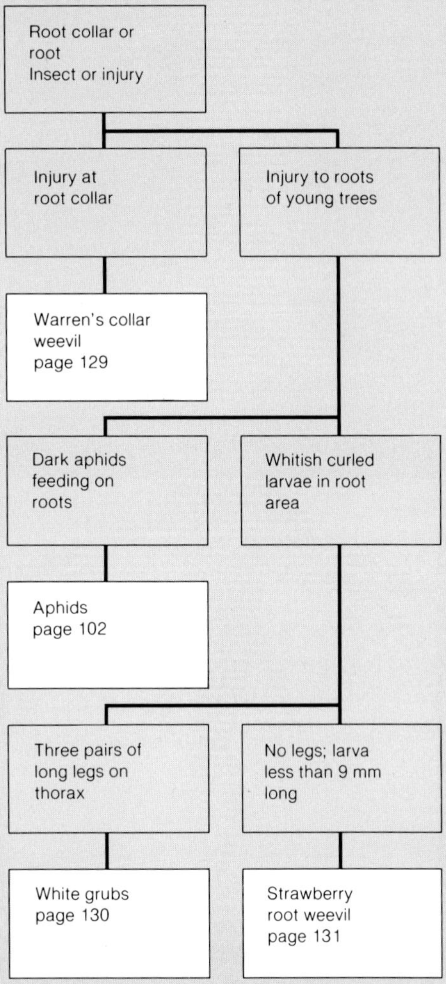

Warren's collar weevil

Girdled root collar

Warren's collar weevil, *Hylobius warreni* Wood, is a common pest of pine and spruce throughout Canada. Although commonly associated with pine, it does cause appreciable damage to native species of spruce both in natural stands and in plantations. Trees growing on wet ground or in deep layers of humus are most susceptible to attack. Tunnels filled with pitch in the root collar region, with larvae feeding in the cambial or inner bark region, are characteristic of attack by this weevil. Small trees are often girdled ☐ and killed whereas feeding on larger trees permits the entrance of wood rots into the wounds, as a result of which the trees are susceptible to wind breakage.

The adults live 2 or more years, laying eggs each year. Since 2 years are required to complete a life cycle, both larvae and adults can be found during the winter. The pupal stage, which is short, occurs in June. The adults, which are flightless, have an extended emergence period.

Large-scale control programs have not been attempted in natural stands because of the scattered nature of the damage. Because the weevil requires a damp site with a deep humus layer to complete development, such sites in plantations should be avoided. This insect has not been a problem on ornamental trees.

White grubs

Larva

White grubs are the larvae of June beetles of the genus *Phyllophaga* and the related genus *Serica*. There are more than 100 different kinds throughout North America. The larvae normally feed on the roots of grasses and other plants but will also feed readily on the tender roots of coniferous seedlings and transplants when these are planted in sod or on old field sites. The various species of white grubs require 2 to 5 years to complete their life cycle, depending on species and location. The larvae live in the soil and are up to 30 mm long. They have a brownish head and cream-coloured body and are usually found in a curled pose □ . The adults are heavy-bodied beetles, light to dark brown in colour. They emerge from soil in spring or summer and feed on leaves of various deciduous trees.

Seedlings planted in areas where white grubs are common should have their roots treated with a stomach poison to protect them until they become established.

Strawberry root weevil

Adult

Damaged tree

The strawberry root weevil, *Otiorhynchus (=Brachyrhinus) ovatus* Linnaeus, occurs across Canada and the northern United States. It is primarily a pest of strawberry plants, but occasionally the feeding of the larvae causes serious injury to seedlings and young transplants in plantations and nurseries.

The weevil overwinters as a larva deep in the soil or as an adult under stones or in other sheltered places. The larvae feed on the roots, weakening or killing small trees. ☐ . They are whitish, C-shaped, and about 8 mm long when full grown. The larvae change to pupae in the upper layers of the soil in the spring. The pupae change to adults ☐ in spring or summer and join the adults of the previous summer to feed on leaves and fruit. The adults do not have functional wings and cannot fly. There are no males and reproduction occurs asexually. The eggs are laid at random in the soil around the food plants in summer and, on hatching, the larvae feed on the roots until late fall, when they hibernate. The adults are often a nuisance in recently built suburban homes in summer and fall when they seek hibernating sites there.

Dipping the roots of seedlings in an insecticide suitable for soil insects and with a long residual life should prevent damage to seedlings. For established young trees a soil insecticide should be raked into the soil in late June.

Flower or cone

*The flower referred to is the male or staminate flower of spruce, fir or hemlock.

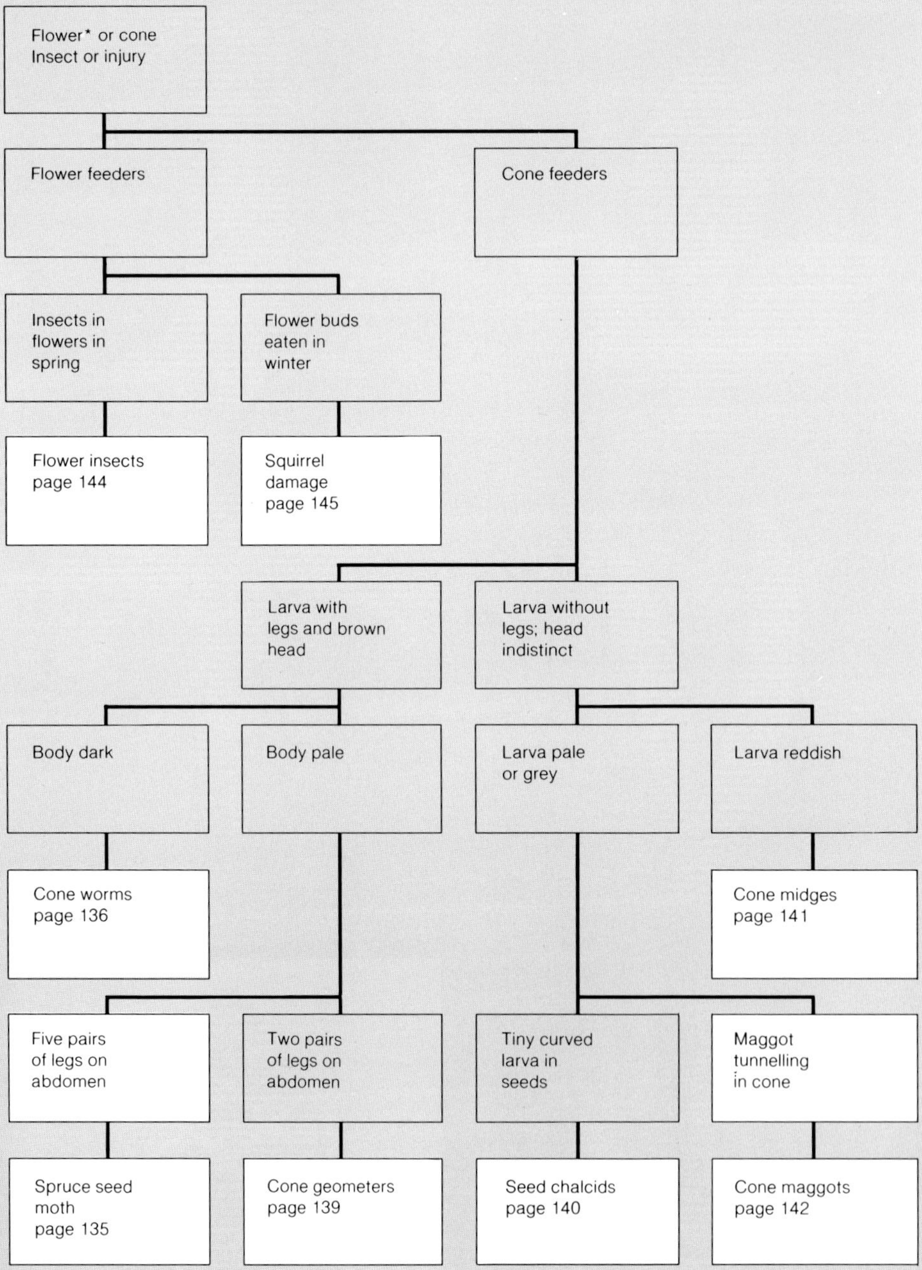

Spruce seed moth

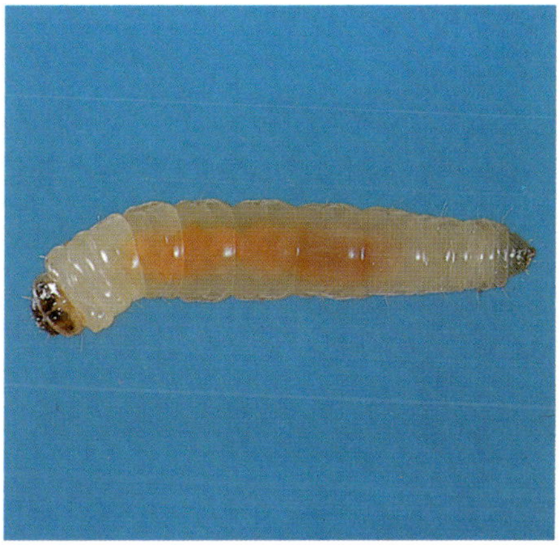

Larva

Larva feeding in cone

The spruce seed moth, *Cydia youngana* (Kearfott), is found across Canada and the northern United States. It feeds in the cones of various spruces, and severe damage to white spruce cones has been reported from central and western Canada.

In Ontario the moths fly in May, about the time the spruce pollen is shed, and lay a single egg per cone if cones are plentiful. On hatching, the young larva tunnels in the cone scales initially and then moves deeper to feed on the developing seeds. In late June the larva ☐ makes a narrow tunnel down the axis of the cone ☐ from which it feeds. When full grown in the fall, and about 10 mm long, it hibernates in the central tunnel in the cone. Larvae change to pupae in the spring and adult moths emerge from the pupal cases in about 18 days.

Some of the larvae do not change to pupae the first spring but remain dormant for 1 or more years. It would appear that the number of adults present in any year correlates with the number of cones.

Control of the seed moth is usually necessary only in seed orchards and is best handled by specialists.

Cone worms

Spruce coneworm

Fir coneworm

Spruce budworm

• Mature larva dark with cinnamon-coloured stripes ......Spruce coneworm
• Mature larva reddish purple without distinct stripes..................Fir coneworm
• Mature larva brown with rows of small pale spots........Spruce budworm, page 55

The **spruce coneworm,** *Dioryctria reniculelloides* Mutuura & Munroe, probably occurs throughout most of the range of spruce on this continent. East of the Rocky Mountains it feeds on the new foliage ☐ and cones of spruce ☐ and often balsam fir. This coneworm is occasionally abundant, often in conjunction with epidemics of the spruce budworm. When the new foliage has been consumed by these two rival species, the coneworm will often eat the budworm or its pupa.

Spruce coneworms overwinter as tiny larvae in silken shelters on the tree. In early spring they tunnel into needles located near flower or leaf buds. As spring growth commences, they move to feed in elongating buds or cones. Feeding is completed in June or July. The full-grown larvae, about 17 mm long, change to pupae in or near feeding sites and adult moths emerge in about 15 days. Eggs are laid in a variety of sheltered niches in the tree crown and hatch in about 11 days. The young larvae go into hibernation without feeding.

The **fir coneworm,** *Dioryctria abietivorella* (Grote), is also widely distributed throughout the continent. It feeds in a variety of sites on many conifers and is regularly found in cones of spruce ☐ and balsam fir, but not in high numbers. On the other hand, it has been reported as causing serious injury to Douglas-fir cones in British Columbia. There is apparently one generation each year in Canada but with a considerable overlap of stages so that larvae may be found in cones from spring to fall. The full-grown larva is about 20 mm long.

Spruce coneworm shoot damage

For control of these species in seed orchards the appropriate forest research centre should be consulted.

Spruce coneworm damage

Fir coneworm damage and an undamaged cone

Cone geometers

Eupithecia mutata larva

The cone geometers, *Eupithecia albicapitata* Packard and *Eupithecia mutata* Pearsall, are found occasionally in the cones of spruce, fir and hemlock across Canada and in the northern United States. Apparently they overwinter as pupae in the soil and the adults emerge in June and July. The larvae are usually found singly, burrowing in the cones in July and August. The mature larva ☐ is about 10 mm long. No extensive feeding damage has been recorded for the two cone geometers and control measures have not been necessary.

Seed chalcids

The **balsam fir seed chalcid,** *Megastigmus specularis* Walley, probably occurs throughout the range of balsam fir. Because the seed is not widely collected, little is known about the damage this chalcid causes except for two reports from Saskatchewan and Manitoba, where heavy damage occurred.

This chalcid overwinters in the larval stage inside a seed on the ground. The larva changes to a pupa in the spring and the adult □ emerges through a hole cut in the seed coat □, about the time that pollen is shed. After mating, the females seek out the young cones and lay their eggs inside the developing seeds by means of the slender, pointed egg tube at the tip of the abdomen. On hatching, the larvae begin feeding, each one reaching maturity on the food contained in a single seed. Full-grown larvae are yellowish white and strongly arched and vary in length from 1.5 to 4 mm.

The **spruce seed chalcid,** *Megastigmus piceae* Rohwer, feeds in the cones of spruce. It is similar to the balsam fir chalcid in appearance, habits, and seasonal occurrence of stages but is apparently much less common.

As with other insects attacking cones, the need for control measures is of concern to a very small number of people and each instance is best treated separately after consultation with specialists.

Adult

Adult emergence hole

Cone midges

Larva in spruce cone

Larvae of the midges *Dasineura* and *Mayetiola* and occasionally those of other genera feed in the cones of spruce, fir and hemlock but apparently have not caused serious injury to seeds. The common species on spruce overwinter as larvae inside cocoons in the previous year's cones. The larvae change to pupae and subsequently to adults in early spring. The two-winged, long-legged females lay their eggs on the young cones. On hatching, the larvae □ of different species feed in different parts of the cone. In spruce cones, for example, larvae of one midge species feed in the developing seeds □ , and those of another feed in a small swelling in the cone scales, while larvae of a third species are usually found in a cavity at the base of a scale. Full-grown midge larvae are about 3 mm long and pink to orange. Some larvae remain dormant in the cocoon over two winters before changing to pupae and subsequently to adults.

Control operations, if necessary in seed orchards, are best left to specialists.

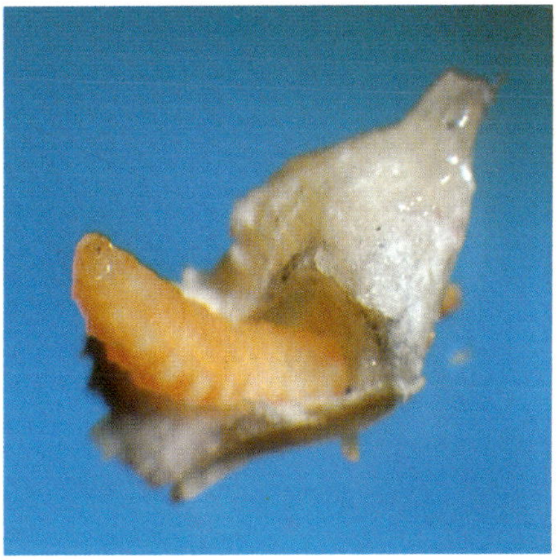

Larva in balsam fir seed

Cone maggots

Hylemya abietis

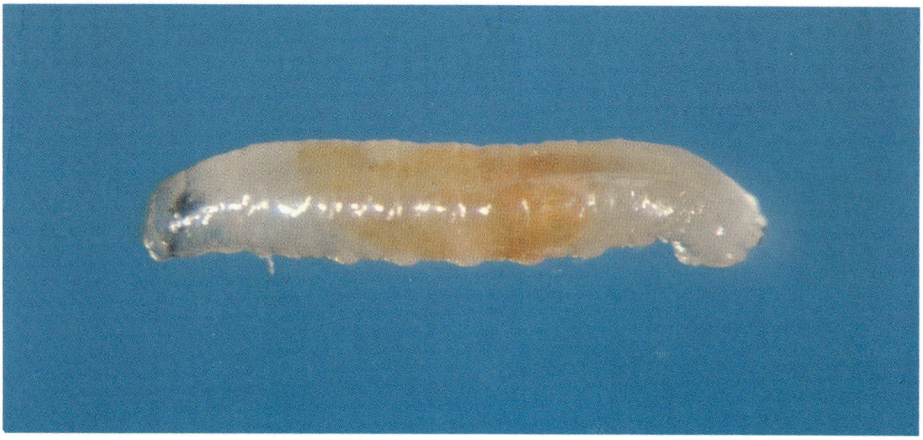

Earomyia aterrima

Although this group of insects has not been studied extensively in eastern Canada, three species have been identified, one on spruce and two on balsam fir. The **spruce cone maggot,** *Hylemya anthracina* (Czerny), is a common pest of spruce. It spends the winter as a pupa in the soil and the adults emerge in spring when the pollen is being shed. The female lays her eggs between the cone scales. On hatching, the larvae tunnel spirally around the cone axis. When full grown in June or July and about 6 mm long, they tunnel out of the cone and drop to the ground, where they change to pupae.

A closely related species, *Hylemya abietis* Huckett, tunnels in the cones of balsam fir. The white larvae ☐ feed in spring and early summer and their circular exit holes ☐ are found in damaged cones later in the season.

A second species in cones of balsam fir is *Earomyia aterrima* (Malloch). In Ontario the larvae ☐ have been found in cones in July and August. The mature larva is about 7 mm long.

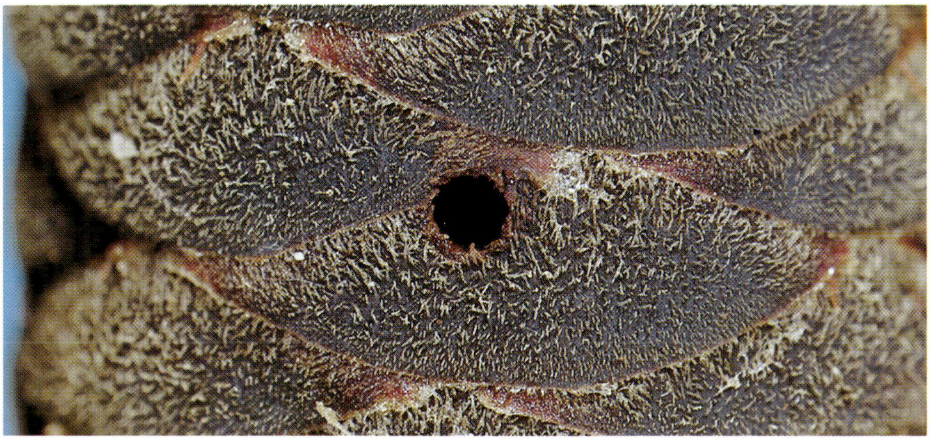

Larval exit hole in cone

It drops from the cone and changes to a pupa on the ground, where it overwinters.

Control measures are required only in seed orchards where special conditions prevail, and each situation should be separately assessed.

Flower insects

Spruce budworm in flowers

The staminate flowers of spruce, fir and hemlock are small, and even though they have a short duration, often harbour numerous tiny insects that feed on pollen. Perhaps because the pollen from flowers is sufficiently abundant to fertilize the available cones, insect inhabitants seldom attract attention and have not been widely studied. Two kinds of insect found commonly in flowers of balsam fir are shown, but no doubt many others occur from time to time. Winter feeding damage by squirrels is also illustrated.

- Small, usually dark larvae .... Spruce budworm
- Tiny, active, golden-hued adult insects .... Flower thrips
- Flower buds eaten in winter .... Squirrel damage

The **spruce budworm,** *Choristoneura fumiferana* (Clemens), often feeds in the staminate flowers □ in spring. In the fall the tiny dark larvae of the succeeding generation often utilize the old flower cups on balsam fir as a hibernation site. (For more

Adult thrips

information on the spruce budworm see pages 54 to 59.) Other young moth larvae also often feed in the flowers in spring but rarely in numbers.

The **flower thrips** □ , *Chilothrips pini* Hood, is often found in staminate flowers of spruce and balsam fir. The winged adults are only about 1.5 mm long.

The necessity for control operations is doubtful.

Squirrel damage

Squirrel-feeding on flower buds

The red squirrel, *Tamiasciurus hudsonicus* (Erxleben), is a common inhabitant of coniferous forests. Its removal of cone-bearing twigs from pine is well known. Balsam fir as well is often pruned by the squirrel in winter as it clips twigs and feeds on the buds of male flowers □ . The discarded twig tips often litter the snow under large trees. Similar feeding damage to leaf buds of Norway spruce has also been recorded.

Because the amount of loss is unpredictable, no control measures have been attempted.

Undamaged white spruce flowers

Undamaged balsam fir cones, scales and seeds

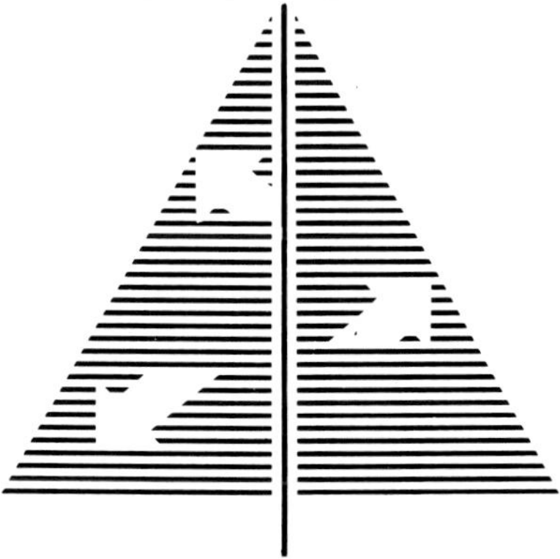

Discoloured foliage

Needles on branches or whole trees sometimes turn brown and no obvious cause of injury can be found. Disease organisms are often directly the cause of discolouration of the outermost needles □ and subsequent loss, usually without tree mortality. However, environmental disturbances such as soil compaction or contamination and soil or water-level changes, especially in housing developments, also cause needle browning throughout the tree □ . In the latter instances death usually occurs, and beetles or borers are frequently found in the trunk of the tree so affected. Needles on recently opened buds may be killed by late spring frosts □ in low-lying areas, but the amount of damage will vary on different trees. Another kind of needle and shoot damage is caused by herbicides □ , which usually cause characteristic distortion of new growth. When the oldest needles □ , those nearest the trunk, turn rusty red in late summer or early fall, there is no cause for alarm, since a natural annual phenomenon is occurring: the trees are losing their oldest needles.

Needle cast of Norway spruce

Herbicide injury

Frost-killed buds

Environmental disturbance

Normal late-summer needle drop

Index

Acknowledgements

Environment Canada

Photography	E.R. Rayner, D.C. Anderson, W.J. Miller
Design and production	Iris Gott. Other illustrative material, G. Guillet and G. Bourgon
Cover photo	E.R. Rayner
Production	A. Lavallée
Other photography and assistance	A.A. Harnden, W.L. Sippell and L.M. Gardiner, Great Lakes Forest Research Centre
	R.S. Forbes, Maritime Forest Research Centre
	P. Benoit, Laurentian Forest Research Centre
	G.J. Smith, Northern Forest Research Centre
	D.M. MacLeod, Insect Pathology Research Institute

Regional Economic Expansion, Canada	Tree Nursery, Indian Head, Sask., photo of spruce spider mite
Ministry of Natural Resources, Ontario	A.G. Gordon for red spruce photo and material

Metric/English conversion scales

Linear

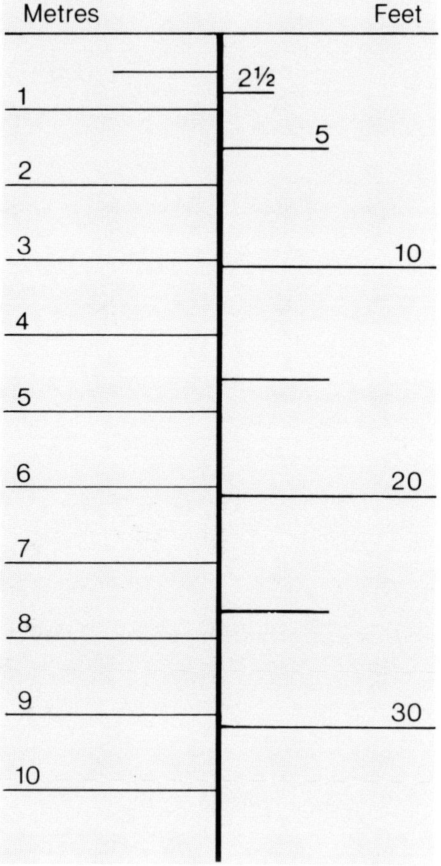

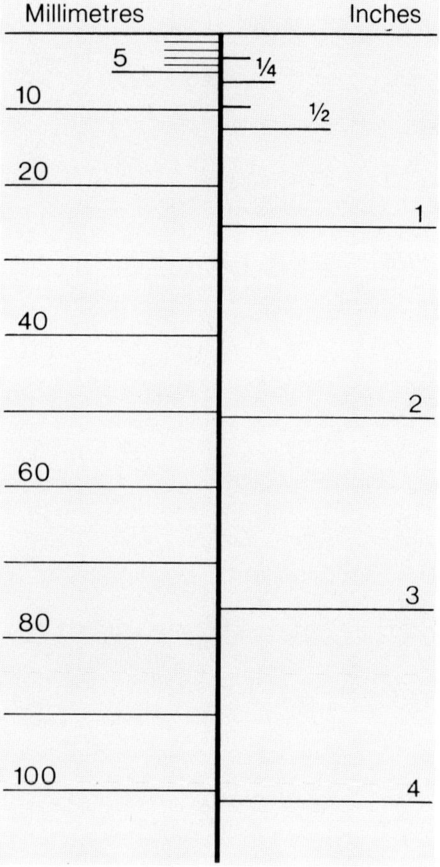

Area

1 cm^2 = 0.155 po^2

1 ha = 2.47 acres

Volume

1 m^3 = 0.276 cord